Tourism, Power and Culture

TOURISM AND CULTURAL CHANGE
Series Editors: Professor Mike Robinson, *Centre for Tourism and Cultural Change,* *Leeds Metropolitan University, Leeds, UK* and **Dr Alison Phipps,** *University of Glasgow, Scotland, UK*

Understanding tourism's relationships with culture(s) and vice versa, is of ever-increasing significance in a globalising world. This series will critically examine the dynamic inter-relationships between tourism and culture(s). Theoretical explorations, research-informed analyses, and detailed historical reviews from a variety of disciplinary perspectives are invited to consider such relationships.

Full details of all the books in this series and of all our other publications can be found on http://www.channelviewpublications.com, or by writing to Channel View Publications, St Nicholas House, 31-34 High Street, Bristol BS1 2AW, UK.

TOURISM AND CULTURAL CHANGE
Series Editors: Professor Mike Robinson, *Centre for Tourism and Cultural Change, Leeds Metropolitan University, Leeds, UK* and Dr Alison Phipps, *University of Glasgow, Scotland, UK*

Tourism, Power and Culture
Anthropological Insights

Edited by
Donald V.L. Macleod and James G. Carrier

CHANNEL VIEW PUBLICATIONS
Bristol • Buffalo • Toronto

Library of Congress Cataloging in Publication Data
A catalog record for this book is available from the Library of Congress.
Tourism, Power and Culture: Anthropological Insights/Edited by Donald V.L. Macleod and James G. Carrier.
Tourism and Cultural Change
Includes bibliographical references and index.
1. Tourism. 2. Tourism--Anthropological aspects 3. Tourism--Social aspects.
I. Macleod, Donald V. L. II. Carrier, James G.
G155.A1T592433 2010
306.4'819–dc22 2009048872

British Library Cataloguing in Publication Data
A catalogue entry for this book is available from the British Library.

ISBN-13: 978-1-84541-125-1 (hbk)
ISBN-13: 978-1-84541-124-4 (pbk)

Channel View Publications
UK: St Nicholas House, 31-34 High Street, Bristol BS1 2AW, UK.
USA: UTP, 2250 Military Road, Tonawanda, NY 14150, USA.
Canada: UTP, 5201 Dufferin Street, North York, Ontario M3H 5T8, Canada.

The policy of Multilingual Matters/Channel View Publications is to use papers that are natural, renewable and recyclable products, made from wood grown in sustainable forests. In the manufacturing process of our books, and to further support our policy, preference is given to printers that have FSC and PEFC Chain of Custody certification. The FSC and/or PEFC logos will appear on those books where full certification has been granted to the printer concerned.

Typeset by Datapage International Ltd.
Printed and bound in Great Britain by Short Run Press Ltd.

Contents

Contributors

Elena Calvo-González received her PhD in Anthropology from The University of Manchester, UK. Since then, she has been researching, as a Postdoctoral Fellow at the Federal University of Bahia, Brazil, issues of embodiment and race with regard to public health policies and medical technologies.

James G. Carrier began studying tourism, environmental conservation and economy in Jamaica and the Caribbean in the mid-1990s. He has supervised or co-supervised projects dealing with these topics in Montego Bay, Negril and Port Antonio, all in Jamaica. He is currently Senior Research Associate at Oxford Brookes University, and Adjunct Professor of Anthropology at the University of Indiana.

Luciana Duccini received her PhD in Social Anthropology at the Federal University of Bahia, Brazil. She is currently a lecturer at Universidade Federal do Vale do São Francisco. Her research interests include African-Brazilian religion and its relationships with identity processes, social class and tourism.

C. Michael Hall was a Professor at the University of Canterbury, New Zealand, at the time of writing. He is also a Docent in the Department of Geography, University of Oulu, Finland and a Visiting Professor, Baltic Business School, University of Kalmar, Sweden. Co-editor of *Current Issues in Tourism*, he has published widely in the fields of tourism, regional studies, gastronomy, environmental history and environmental change.

Michael Hitchcock is Deputy Dean (Research and External Relations) at the University of Chichester and formerly Director of the International Institute for Culture, Tourism and Development at London Metropolitan University. While teaching at Hull University in the Department of Sociology and Anthropology and the Centre for South-East Asian Studies, he was the co-editor of *Tourism in South-East Asia* (Routledge, 1993).

Teresa Holmes is an Associate Professor of Anthropology at York University in Toronto (Ontario), Canada. Her teaching and research interests include tourism studies, colonial and postcolonial culture, historical anthropology, and critical kinship studies. Her chapter in this volume is part of an ongoing project on issues of ethnic citizenship and tourism in Belize.

Charlotte Joy is an ESRC-funded Postdoctoral Fellow at the Institute of Archaeology and Anthropology, University of Cambridge. During her doctoral research, she carried out 10 months of fieldwork in Djenné, Mali and spent two months at UNESCO's Intangible Heritage Department in Paris. She is specialising in developing a comparative ethnographic approach to the study of cultural heritage politics and its relationship to development issues.

Donald V.L. Macleod trained in anthropology at Oxford University and is a Senior Lecturer at the University of Glasgow where he has run two research centres. He has researched in the Caribbean, the Canary Islands and Scotland, and published widely on tourism impacts, cultural change, globalisation, identity, sustainable tourism development and heritage. His books include *Tourism, Globalisation and Cultural Change* (2004), *Niche Tourism in Question* (2003, editor), *Tourists and Tourism* (1997, co-editor).

I Nyoman Darma Putra is currently a Postdoctoral Research Fellow at the School of Languages and Comparative Cultural Studies, University of Queensland. He is a lecturer at the Indonesian Department, Faculty of Letters, University of Udayana, Bali. With Michael Hitchcock, he published *Tourism, Development and Terrorism in Bali* (Ashgate, 2007).

Gunilla Sommer studied anthropology at the University of Copenhagen and did field work for her master thesis in a Bolivian rain forest community working with ecotourism. After completing her studies, she did field work on tourism in Jamaica as a research assistant and later she worked in the tourism business in Bolivia and Egypt.

Veronica Strang is an environmental anthropologist at the University of Auckland. She has written extensively on water, land and resource issues in Australia and the UK, and is the author of *The Meaning of Water* (Berg, 2004); and *Gardening the World: Agency, Identity, and the Ownership of Water* (Berghahn, 2009).

Dimitrios Theodossopoulos teaches anthropology at the University of Bristol. His earlier work examined people-wildlife conflicts and indigenous perceptions of the environment. He is currently working on ethnic stereotypes, indigeneity, authenticity and the politics of cultural representation in Panama and South-East Europe. He is the author of *Troubles with Turtles: Cultural Understandings of the Environment on a Greek Island* (Berghahn, 2003), and editor of *When Greeks Think about Turks: The View from Anthropology* (Routledge, 2006) and *United in Discontent: Local Responses to Cosmopolitanism and Globalization* (Berghahn, 2009).

Preface

The two of us organised a panel for the annual meeting of the Association of Social Anthropologists of the UK and Commonwealth (the ASA), held in March 2007 at London Metropolitan University. The theme of the meeting was 'Thinking through tourism'. Both of us were frustrated by the limited scope of much of the work that was being done on tourism in anthropology and elsewhere. We thought that the work was often formulaic and mechanical and too little concerned with power and what was going on more widely in the tourist destinations' countries. We hoped that we could recruit people concerned with a set of questions that deserved more attention. Those questions revolved around the interplay between tourism and political-economic relations within host countries, influenced by the cultural context. We selected the best-suited of the papers from that panel, and the result is *Tourism, Power and Culture: Anthropological Insights*.

These papers are intriguing and revealing. In them, tourists as individuals visiting a tourist destination are not our main concern. Instead, they are more important in the aggregate, as the people who, collectively, make up consumer demand and are catered to by the tourism sector. Also in the papers in this volume, the encounter between hosts and guests is important only in so far as it influences and is influenced by national and international interests and processes that extend well beyond the immediate places of the tourist destination and the immediate nature of the tourist encounter.

The papers illustrate a growing body of anthropological work on tourism, and they are part of an increasing interest in the discipline in power in its more straightforward forms. Importantly, they show the ways that a broader perspective on tourism helps illuminate processes and relations that are of serious relevance for tourist destinations and host societies, but that are difficult to discern if we restrict our attention to tourists and the encounter between hosts and guests. And finally, they illustrate one of the central strengths of anthropology: training in the discipline routinely involves a year or so of participant observation in a field site, coupled with attention to people and processes that are important for the site but are located elsewhere. The result is not just a close knowledge of what goes on in that site and a deep understanding of the host community's culture. In addition, such protracted fieldwork increases the chance that the researcher will come across the unexpected, things that could not have been anticipated but are relevant for understanding the people and events that they study.

A volume like this is a reflection of the influence and efforts of many people. We want to thank Tom Selwyn, who suggested that the two of us combine our interests and organise a joint panel at that ASA meeting. We also want to thank the contributors, who cheerfully tolerated our editorial suggestions, accepting the good ones and politely ignoring the bad. Finally, we want to thank Michael Hall, who was prepared to take time from his own schedule and devote it to contributing a concluding chapter to this volume.

Donald V.L. Macleod and James G. Carrier

Prologue

Chapter 1

Tourism, Power and Culture: Insights from Anthropology

DONALD V.L. MACLEOD and JAMES G. CARRIER

Introduction

Tourism is inextricably linked with power and culture in many ways, including the relationship between those countries providing the tourists and the destination 'host' countries; the interaction between the tourists themselves and the indigenous population of the destination country; the tourism industry structures involving multinational companies; and the political interests at every level with a concern for the economy. Culture, in its broadest sense, is the framework within which tourism takes place, as all people have a cultural background and much of tourism involves travelling into a different cultural environment. Culture in a narrower sense is something that may be an attraction for the tourists, whether it is in the form of a museum, architecture, music or religious ritual. Culture also helps determine what the tourist wants to do, as a result of formal and informal education, values, family background and cultural mores.

Anthropologists are in an exceptionally good position to provide insights into tourism, power and culture for several reasons: (1) they study culture in its broadest sense, including politics, cosmology, economics, kinship, customs and material culture, and they take an international approach that involves cross-cultural comparisons and looking at all types and levels of society. Generally, a holistic approach is taken, seeing aspects of social life as interrelated, such as economics and politics, and the topic of interest can move from the general and abstract to the specific and concrete. (2) One of the main research methods involved is 'participant observation', which includes fieldwork, often of 12 months or more for a doctoral project. There follows a life-long interest in the area and frequent return visits for the professional

3

anthropologist. Such an experience gives the anthropologist a deep and detailed understanding of the people and culture studied and the ability to make nuanced and well-informed assessments of developments, such as the arrival of tourists to an area. (3) By learning the local language and interacting intensively with the local people, anthropologists are in a position to interpret their worldview and begin to understand the meaning of their actions and statements – the 'emic' perspective, as opposed to an objective 'etic' view. (4) Because of the long-term view that anthropologists have of their subject matter, a tourism destination site for example, they are able to note changes over the period of time and also record the perceptions of indigenous people and the visitors.[1]

This book provides insights into tourism, power and culture from anthropologists who give detailed examples from locations worldwide: Queensland, Australia; Mali; The Canary Islands, Spain; The Dominican Republic; Bali, Indonesia; Brazil; Jamaica; Belize; and Panama. There is a similar broad reach in the subject matter of the chapters, including: struggles between farmers and tourism groups over water as a resource; disagreements between United Nations officials and villagers over a World Heritage Site and its building materials; the use of cultural heritage by officialdom and grass-roots populations; the use of terrorism and the need for security as a resource for a successful destination; culture as a resource to be exploited by indigenous people; ethnic diversity and its promotion as an attraction; the notion of blackness and its relation to tourism development; the way tourism agencies construct other groups for their own gain. These diverse topics are united in their involvement with power in some form or capacity. Likewise, the concept and concrete reality of culture pervades each case study.

By focusing on tourism, power and culture, this collection of work by social scientists, most of whom have studied anthropology, is unique. It provides valuable material for those interested in the study of tourism, as well as anthropology, and, of course, it is of value to those interested in power and culture. The disciplines of Tourism Studies and Social (or Cultural) Anthropology are clearly linked areas of academic endeavour that match the subject matter covered by this book. Equally, the intellectual focus is on these disciplines, where relevant sources and references are discussed below. This volume[2] deliberately builds on work already completed in the disciplines concerning tourism and power (the central issues), and provides new material in the form of case studies, arguments and ideas.

Understanding Power

Anthropological approaches

Power has been used as a concept by anthropologists in a manner that accepts a wide variety of interpretations and manifestations. The approach in this volume follows other anthropologically oriented works on power, notably Fogelson and Adams (1977) and Cheater (1999a), which deal with power in numerous guises using many different cultural and geographic settings; furthermore, this volume is not prescriptive concerning the definitions employed by contributory writers. This section will review how the concept of power has been dealt with by anthropologists, in so far as it is relevant to the theme of this book, tourism.

In his book examining power and organisations in complex society, Cohen (1976: 23) notes the strong linkage between economic and political interests: 'they continually exert pressure on the state and the state continually exerts pressure on them'. This observation is particularly appropriate to the tourism industry, with its ability to influence powerful bodies, including government organisations. Cohen (1976: 23) also states: 'On a high level of abstraction all social relationships have their aspect of power. Power, as many scholars have pointed out, does not exist in the abstract but always inheres in social relationships'. He adds: 'on the other hand, relations of power are aspects of social relationships' (Cohen, 1976: 34). This is useful to bear in mind when considering the interplay between individual people, groups and organisations where tourism is influencing particular sectors. Moreover, it indicates the fundamental location of power in all human social circumstances.

The need to explore and analyse power within human societies was addressed by Fogelson and Adams (1977) in an edited collection that sought to use the notion of power in its broadest sense and apply it in a manner comprehended by cultures from Asia, Oceania and the New World. In a summary chapter for the substantial volume, Colson (1977: 376) drew attention to a type of power that exists in its own right (that possessed by Shamans for example), as well as the power that exists as a relationship between people with different resources. She remarks of many examples within the book that: 'They saw power not as an entity in itself, but as the ability to bend others to one's ends. To fail is to be powerless. To have no resources with which to trade is to be powerless' (Colson, 1977: 376). The importance of resources is well recognised within the tourism industry in which comparative advantage based on natural and cultural assets is crucial for a competing destination. The

ownership of resources is an issue debated when considering development in its broadest sense, whether we include dependency situations
(cf. Britton, 1989; Harrison, 2001; Lea, 1988) or the leakage from an all-
inclusive resort run by expatriate management and owned by a foreign
company (cf. Lea, 1988). Contributors considering resource allocation
and competition in this volume include Strang, who examines the
struggles over water; Joy, who examines the vernacular architecture
that has become a World Heritage Site and subject of contestation;
Macleod, who considers cultural heritage as a resource; Theodossopoulos, who looks at how an indigenous group are able to use their culture as
a resource; and Calvo-Gonzáles and Duccini, who discuss how ethnicity
is being used by government organisations as a promotional attraction.

Pursuing a more abstract theme in the aforementioned volume,
Adams (1977) sought to develop a framework that could relate the
various phenomena referred to by the contributors, a goal that reflects
the anthropological tradition of seeking universals through comparison.

> If we take "power" to refer to the ability of a person or a social unit to
> influence the conduct and decision-making of another through the
> control over energetic forms in the latter's environment (in the
> broadest sense of that term), we may then differentiate two kinds of
> power relations: *independent* and *dependent*. Among the second we
> can further differentiate power *granting allocation*, and *delegation*.
> Independent power is the relation of dominance based upon the
> direct abilities and controls of an individual or social unit. Power is
> dependent when one controller gives (although it would be more
> accurate to say "lends") another the right to make decisions for him.
> (Adams, 1977: 388)

He summarises by stating: 'Power then is a relational quality that exists
contingent on controls that can be exercised over elements in the external
world. As such it is seen ethnographically to exist differentially and
independently for all men and may be extended to many things' (Adams,
1977: 395). Adams has provided a very useful framework and working
definition with which it might be possible to understand social phenomena in the field and applied to situations involving tourism. Macleod (this
volume) uses the framework as a springboard for further analysis and
develops it in relation to analysing cultural heritage and tourism.

A more recent edited collection (Cheater, 1999a), again focusing on the
anthropology of power, reflected the contemporary interest in empowerment, and many of its chapters looked at fundamentally political
situations with a wide range of topics and locations embracing gender,

race, post-colonialism, multiculturalism, development and state control. Max Weber's concept of legitimised power, with its distinction between power (as the ability to elicit compliance against resistance) and authority (as the right to expect compliance), forms a theme for some contributors. Michel Foucault's distinguishing between central 'regulated and legitimate forms of power' and 'capillary' power at the 'extremities' is also influential among contributors (Cheater, 1999b: 2). A modernist view is discussed by James:

> The concept of power used to be based, in one way or another, upon the ownership and productive use of resources, or upon the control of people through historically established political formations in which legitimacy was invented and specifically located, often though it might be contested and usurped. (James, 1999: 14)

Such a concept of power would fit with many of the concerns and examples within this volume; however, this does not deny a postmodernist recognition of the varieties, subtleties and complexities of power (which is not simply a one-way process), which is addressed by the focus on empowerment in Cheater's volume.

Politics is addressed straightforwardly in a monograph on power by Gledhill (1994) and is the theme in anthropology that has, understandably, attracted most work concerned with the idea of power (Gellner, 1983; Wolf, 2001). Gledhill (1994: 148) refers to Foucault: 'Foucault sees power relations as present in all social relationships, permeating society in a capillary way rather than coming down from a single centre of control such as the state'. Foucault has been particularly influential among some tourism scholars (e.g. Urry, 1990) as well as anthropologists, and his notion of the professional gaze has proved a very strong analogy for understanding tourists and the power of the media and education as employed by Urry (1990) with the notion of the 'tourist gaze'. The need to comprehend the viewpoint of the tourist is of increasing importance within the industry and for the destinations, and this desire is similar to the goal of anthropologists to comprehend the worldview of the people they study.

Anthropologists have treated power with a relaxed and open attitude towards definitions, whilst retaining a desire to remain comparative with their broad-ranging examples. Theorists, including Marx, Weber, Foucault and Bourdieu, remain useful references for many scholars and the distinctly political continues to be the favoured branch of enquiry. Power as a concept or a central organising theme has not yet been applied to tourism by anthropologists in a single collection, although issues such as imperialism (Nash, 1989), development (Harrison, 2001; Smith & Eadington, 1992), as well as political economy (Bianchi, 2000) have been examined by individual scholars in numerous publications.

Nash (1989) deals with imperialism in a theoretical overview introducing the influential edited volume *Hosts and Guests: The Anthropology of Tourism*. His chapter entitled 'Tourism as a Form of Imperialism' utilises a broad definition of the term:

> At the most general level, theories of imperialism refer to the expansion of a society's interests abroad. Those interests – whether economic, political, military, religious, or some other – are imposed on or adopted by an alien society, and evolving intersocietal transactions, marked by the ebb and flow of power, are established. (Nash, 1989: 38)

However, in his monograph, *Anthropology of Tourism* (Nash, 1996), the word 'imperialism' does not even appear in the index, and he writes:

> So, unless the concept is carefully considered and applied, the notion of tourism as a form of imperialism may sometimes be inapplicable in tourism destination areas. It is only that anthropologists' pervading concerns with the underdogs of the world has made it seem so. (Nash, 1996: 28)

Nash chooses to focus on issues such as acculturation, development, policy and sustainable tourism, and he employs the concept of power throughout the book. Nevertheless, if we consider imperialism in its broadest sense of influence relating to ideas and economic or political interest, then work examining development in its manifold forms also deserves mention, as power relations are inextricably linked into these processes.

A seminal collection focusing on developing countries (including East African nations) and their experiences of tourism being used as a tool for development was the volume edited by De Kadt (1980), which took a critical approach to the impact of tourism. This was among the earliest publications that began to seriously challenge and criticise the results of tourism development, drawing attention to negative repercussions. Other such approaches include Britton (1989) and Lea (1988), in which the experience of the local populations was made explicit, as well as placing the tourism development process into the wider political economy of the global system. Implicit in these studies is the asymmetrical power relationship between the developed countries, which send tourists and where tourism firms are based, and the recipient developing countries, which become service economies, dependent on the desires, actions and objectives of other countries.

More recently, Wood (1993) examines the ideology of development and Richter (1993) considers the elite-driven tourism policy in a volume focusing on tourism in South East Asia. Development in its numerous forms, including imperialism, modernisation, dependency and sustainability has become a dominant paradigm in tourism research and economic activity worldwide. Anthropologists are particularly well placed to articulate the grass-roots experience of the indigenous populations, and are also able to experience and communicate the complex dynamics of cultural differences and clashes arising because of the meeting of numerous perspectives, ideologies and worldviews held by distinct groups of people.

Tourism studies and power

As with anthropology, there is a similar pattern of scholarly interest focusing on politics (often government and policy making) and development discernable within the multidisciplinary work of Tourism Studies (e.g. Burns & Novelli, 2006; Elliot, 2004; C.M. Hall, 1994; D. Hall, 2004; Richter, 1989; Telfer & Sharpley, 2008; Hall, 2007; Robinson & Boniface, 1999). Very few books have tackled the concept of power as a theme in its broadest theoretical conceptualisation, but one major exception is edited by Church and Coles (2007a: 2), who state clearly in the opening chapter: 'We wish to place constructs of power more firmly at the centre of the agenda of critical tourism research'. Their edited collection addresses a wide diversity of intersections between tourism and power, including manifestations, expressions and articulations. They recognise that discourses on power are related to specific contexts in which they are developed, including a variety of disciplinary positions, and they note, after Hannam (2002), that the emphasis on power in tourism research has been only very recent and that there has been a shift towards investigating social and cultural relations of power. When considering approaches to power in discourses in tourism, they present the following trends: (1) a plurality of approaches in understanding power; (2) the contestability of power as a concept; (3) disagreements over language used to discuss power; (4) the relevance of debate over the use of the power discourse and why concepts of power are analytically valuable (Church & Coles, 2007: 9).

In their opening chapter, Coles and Church (2007: 14) review the place of power and its definitions in relation to Tourism Studies and relevant work in great detail, drawing on a substantial range of scholastic material, for example references are made to Max Weber, who viewed power (macht) as 'the probability that one actor within a social relationship will be in a position to carry out his own will despite resistance, regardless of the basis on which the probability rests'. Steven Lukes is also referred to frequently:

> Lukes (2005: 30, 37) concluded that: "A exercises power over B when A affects B in a manner contrary to B's interests". Latent power operates ideologically to shape people's thoughts and wishes so that (otherwise apparent) differences of interest are obviated. (Coles & Church, 2007: 21)

Additionally, they use Anthony Giddens as an example of someone who gives a multidirectional version of power: 'Agents can utilise causal power, which, according to Giddens (1984), can both challenge as well as maintain structures. Thus, power is emancipatory and not simply constraining' (Coles & Church, 2007: 22). Such a perspective relates to the idea of empowerment, and this is defined by them:

> Empowerment is a process of enabling, ascribing or authorising the relatively powerless with greater power and it has been a central part of the sustainable development agenda that talks about the plight of local people, especially in developing countries. (Coles & Church, 2007: 272)

This has a bearing on the huge drive for 'sustainable tourism', which is almost universal and used as a concept underpinning many tourism development agendas (see Ioannides *et al.*, 2001; Mowforth & Munt, 1998; Ritchie & Crouch, 2003; Telfer & Sharpley, 2008; Wahal & Pigrim, 1997).

Another view on power that has direct relevance for this collection is the conceptualising of power in terms of 'modalities' by Westwood (2002: 135), who delineates the following: repression and coercion; power as constraint; hegemony and counter-hegemony; manipulator and strategy; power/knowledge; discipline and governance; and seduction and resistance. Westwood views power as being constitutive of social relations. Some of these modalities are highly appropriate to the tourism industry and the experiences of people, both inhabitants and visitors at destinations, notably repression and coercion, manipulation and strategy, seduction and resistance. Chapters in this volume dealing with these modalities are those authored by Macleod, Joy, Strang, and Sommer and Carrier.

This section on understanding power has concentrated on anthropological research in accordance with the focus of the book. It has also given a brief outline of some relevant work from tourism studies, but a full review is beyond the scope of this volume. Following the direction of anthropological interest, we now look at the concept of 'culture' and consider the special importance that it has for anthropological research, and the insights that can be offered into understanding the relationship between tourism and power.

Culture

Culture has interested anthropologists since the beginning of the discipline in the last third of the 19th century (see Barnard, 2000; Schultz & Lavenda, 1995), and, of course, approaches to culture have varied over that period. As well, they vary among the different national approaches to anthropology, most notably between the common British concern with the social side of human existence and the common American focus on the cultural side. But whether American or British, German or French, anthropologists generally have been concerned with human culture and its relationship with the ways that human societies are organised and operate. This concern is, in some ways, the hallmark of the discipline.

Given that so many anthropologists of so many different national traditions have been concerned with culture for so long, it is not surprising that the term has different meanings in different contexts. One of the oldest definitions is still one of the best. Well over a century ago, the English anthropologist, Tylor, defined culture as

> that complex whole which includes knowledge, belief, art, morals, laws, customs, and any other capabilities and habits acquired by man as a member of society. (Tylor, 1871)

A century later, an American anthropologist produced a definition that reflected the growing concern throughout the discipline with meaning:

> Believing with Max Weber that man is an animal suspended in webs of significance he himself has spun, I take culture to be those webs, and the analysis of it to be therefore not an experimental science in search of law, but an interpretive one in search of meaning. (Geertz, 1973: 5)

A decade after that, Adam Kuper, born and raised in South Africa but long living in the UK, put forward a simpler and subtly different definition:

> The way of life of a people. (Kuper & Kuper, 1985)

These definitions indicate the range of ways that people in the discipline have thought about culture. Underlying this range, however, is a fairly basic distinction, one that is important for those concerned with tourism and tourists. That distinction is complex, but the core of it can be captured by distinguishing between, firstly, culture as a set of attributes that individuals acquire as a result of growing up among a particular set of people and, secondly, culture as a thing that distinguishes groups from each other.

The first sense of 'culture' is expressed nicely in Tylor's definition, with its invocation of 'capabilities and habits acquired' by a person as a result of living in a society. Speaking American English, driving on the right-hand side of the road, writing from left to right using letters developed by the Romans, all are culture. They are capabilities and habits that are acquired by growing up and living in particular places and among particular people: someone growing up in Paris would speak differently, someone growing up in Tokyo would drive differently, someone in Damascus would write differently and someone in Moscow would use different letters.

These examples are fairly straightforward, but 'capabilities and habits' covers much more than just this sort of thing. Thus, in his definition of culture, Geertz points to webs of significance, which is to say ways of interpreting and thinking about things around us. These range from the seemingly simple (where it is appropriate to display one's country's flag and where it is not) to the nuanced and subtle (the degree to which one is seen as the author of one's own fortune and misfortune, and the degree to which one is not).

These cultural elements may be easy to name and to imagine, but they can be difficult to uncover. That is because often they become visible only in the subtle interplay and unexpected activities of daily life. Anthropologists seek to know what they are, to learn the worldview of the people they study, in order to understand and interpret the way these people make sense of and explain their experiences. In this manner, through long-term fieldwork living with a group of people, anthropologists will eventually acquire a rich insight into their behaviour, social structure and intellectual outlook. This leads to an awareness of the cultural context in which different people operate, and hence develop sensitivities towards the reasons why they act as they do.

Elements of culture are not only of interest to anthropologists. They are, after all, an account of how people see and think about the world, and hence an account of how they are likely to respond in different situations. Such an account is useful for anyone who wants to know what leads people to decide, for instance, to spend a vacation travelling rather than sitting at home. Similarly, such an account is useful for anyone who wants to know what leads people to welcome strangers into their towns and cities rather than treating them with indifference, hostility, contempt or fear.

We said that seeing culture as a set of attributes that individuals acquire is only one of the main ways anthropologists have approached the subject. The other way is as a thing that groups have, Kuper's 'way of life of a people'. Such a concern has deep roots, particularly in American anthropology, which has long tended to see people in terms of the groups of which they are members, and has long been interested in discovering and making sense of the culture of those groups (e.g. Benedict, 1947).

This approach to culture is not, however, only found among anthropologists. The people who anthropologists study often approach culture in the same way. This is especially likely to be the case when members of different groups come into contact with each other. When that happens, people are likely to become concerned with the identity of the two groups, what distinguishes 'us' from 'them', which often means their cultures. In the last 20 years or so, identity has been an especially important topic in anthropology, and much work has been done on how people come to see themselves as members of groups. Initially, the groups of interest were national (Gellner, 1983), but increasing interest was paid to groups of other sorts, especially ethnic groups (see Eriksen, 2002).

Like the preceding approach to culture, this one is important for those concerned with tourists and tourism, but in a different way (cf. Abram *et al.*, 1997). While the preceding approach allowed us to consider, for instance, what leads people to travel and to act as they do, this one allows us to consider a factor of growing importance in tourism, the attraction of the novel and the exotic that has been part of what appears to be growing disaffection with mass tourism.

For some time, anthropologists have described the ways that indigenous ethnic groups have used aspects of their culture as a political resource. Some Amazonian Indian groups, for instance, have sought to preserve their lands and livelihoods through sustained campaigns that draw on their distinctive dress, bodily adornment and customs (Conklin, 1997; Conklin & Graham, 1995). Some of these campaigns have been fairly successful in their immediate political aims. As well, they have made these groups more visible internationally, and more attractive to tourists, themselves increasingly interested in the alien and the exotic.

As Richard Wilk (1995) observed, however, simply being alien or exotic is not enough, and here the two broad definitions of culture meet. Wilk argued that the very notion of the exotic is itself an aspect of many people's cultural endowments in Western societies. Most notably, only certain aspects of a group's culture are culturally significant in identifying them as an exotic, alien group. Dress is important, as are styles of cooking, dancing and making music. However, some other aspects are not important, like kinship terminology, property rights and linguistic structures. A group can have unique kinship terminology, property and language, but it is not likely to become identified as alien and exotic if its members dress, cook, dance and make music in ways that are familiar to Westerners. Therefore, for those groups that want to attract outside interest, either as support for political claims or as visitors who will spend money, being different is not enough. Rather, members of those groups need to learn to be different in recognisable ways. As Wilk puts it, they have to learn to be local.

Thus, anthropological work on culture, the focus of the second part of this volume, has identified culture in two different ways, both of which are pertinent for those concerned with tourism. The first of those identifications sees culture as orientations or attributes that people acquire from living in their social group. As we said, these help account for why people want to go on vacation to exotic places, and why people respond to tourists as they do. It is important to recognise, though, that these orientations and attributes are contingent: as the situation in which people lead their lives changes, these orientations and attributes will change as well.

The second identification sees culture as a property of groups, their characteristic ways of life. This approach to culture has a different sort of importance for those concerned with tourism, for characteristic ways of life are the cluster of attributes that can make a set of people an identifiable entity that is attractive to tourists. Unlike the previous identification of culture, this one is likely to be relatively stable. That is because a set of people whose culture, in this second sense, serves their purpose are likely to want to maintain that culture for as long as possible.

As the preceding paragraphs have suggested, culture in its second form feeds back into the issue of power discussed in the first part of this Introduction. As we have noted, anthropologists have described different ways that having an exotic culture of the right sort can be a political or economic resource, and hence can be useful for groups seeking to achieve their goals. There are many such goals, but the ones that have attracted the most attention of anthropologists are political, in the broad sense of the word, as different groups seek different sorts of recognition and support, especially from government. In some cases, as we noted, the right sort of exotic culture can attract attention and support from outsiders, notably in the case of Amazonian Indians in their disputes with national governments. In other cases, that culture can attract tourists, and their interest and wealth can be a resource of the groups in question in their dealings with governments and with other groups.

We want to close this discussion of culture with one further observation about its relationship with power. We have observed that a culture can be a resource in political struggles and can be a source of wealth. In light of this, it should not be surprising that culture itself becomes the focus of political conflict, as different people will seek to control it. These efforts to control a culture can take different forms and revolve around different questions. One such question centres on the control of authenticity. Which person or body has the authority to define what is authentic to that culture, and so deserves to be supported, and what is inauthentic, and so should be denied support? Another such question is representation. Which person or body has the authority to represent the group that bears the culture, in its negotiations with other groups and especially in its negotiations with government? We could go on, but what we have said here is enough to indicate the ways that culture and power are intertwined.

Summary

In summary, this volume presents case studies that illustrate the machinations of power in diverse situations around the world in which understanding the cultural context of tourism-influenced activities is crucial to gaining an insight into the complexity of processes and the variety of perceptions held by differing groups of people. Consequently, the insights of anthropologists gives the readers a rich understanding of the specific examples illustrated in this text, as well as other situations with which they may make comparisons.

This volume is divided into two main parts: the first part, Tourism and the Power Struggle for Resources, deals with the struggle over resources at the tourist destination, where different groups experiencing different levels of power seek control over physical resources, including water, material culture, material heritage and the more abstract state of security against terrorism. The second part, Tourism and Culture: Presentation, Promotion and the Manipulation of Image, deals with the use of culture by different groups for their own specific goals, often in conflict with others: including the use of indigenous culture as a resource; the manipulation of ethnicity and national identity; Black culture and its promotion; and the portrayal of tourists, traders and fishers by the tourism industry. Each part has an introduction, describing the core themes of the chapters and placing them into an embracing comparative framework.

To conclude the volume there is an epilogue: Power in Tourism: Tourism in Power written by Michael Hall, which moves away from the anthropological case study approach and considers the concept of power and its place in tourism studies, as well as its relationship to the tourism industry and the academy. It provides an examination of power, which critiques the concept emphasising its multilayered aspect, looks into non-decision making, and in drawing upon the writing of Lukes (referred to above), offers a critical review of Foucauldian ideas on power. Hall makes a plea for more work on power in tourism studies and a need for understanding the broader socio-cultural context of power, as well as to appreciate its existence in the less obvious experiences: akin to the 'habitus' of Bourdieu, and 'capacity' as described by Lukes. These observations are supplied by examples, such as those of indigenous groups from Australia, New Zealand and Canada, whose political status have relevance to their experiences with tourism development, but are often obscured. Hall turns his attention, eventually, onto the academy and provocatively suggests that there should be more reflection on power relationships within academic institutions as well as other communities observed by scholars. He continues discussions embarked upon by earlier chapters concerning theory and culture, uses examples that have strong anthropological resonance, and causes the reader to reflect on the practices of anthropology and other disciplines concerned with tourism.

A final comment from Coles and Church is of special relevance:

> Given the tourism academy contains many researchers with backgrounds in anthropology, geography and sociology it is still curious that power, a core concept in social sciences generally and more importantly very recently, has not become a more prominent issue in tourism research. (Coles & Church, 2007b: 270).

It is hoped that this volume will help to fill the gap in research and publication.

Notes

1. See the following introductory texts: Burns (1999); Franklin (2003); Lienhardt (1966); Mair (1972); Nash (1996); Smith (1989).
2. The origins of this book lie in the ASA Conference 2007 entitled 'Thinking Through Tourism' during which a panel chaired by Donald Macleod and James G. Carrier considered tourism, political economy and culture. Papers presented at the conference provide the foundation for this collection.

References

Abram, S., Waldren, J. and Macleod, D.V.L. (eds) (1997) *Tourists and Tourism: Identifying with People and Places*. Oxford: Berg.

Adams, R.N. (1977) Power in human societies: A synthesis. In R.D. Fogelson and R.N. Adams (eds) *The Anthropology of Power: Ethnographic Studies From Asia, Oceania, and the New World* (pp. 387–410). New York: Academic Press.

Barnard, A. (2000) *History and Theory in Anthropology*. Cambridge: Cambridge University Press.

Benedict, R. (1947) *The Chrysanthemum and the Sword*. New York: Houghton and Mifflin.

Bianchi, R.V. (2000) The political and socio-cultural relations of world heritage in Garajonay National Park, La Gomera. In M. Robinson, N. Evans, P. Long, R. Sharpley and J. Swarbrooke (eds) *Tourism and Heritage Relationships: Global, National and Local Perspectives*. Sunderland: Business Education Publishers.

Britton, S. (1989) Tourism, dependency and development: A mode of analysis. In T. Singh, H. Theuns and F. Go (eds) *Towards Appropriate Tourism: The Case of Developing Countries* (pp. 93–116). Frankfurt am Main: Verlag Peter Lang GmbH.

Burns, P.M. (1999) *An Introduction to Tourism and Anthropology*. London: Routledge.

Burns, P.M. and Novelli, M. (eds) (2006) *Tourism and Politics: Global Frameworks and Local Realities*. Oxford: Elsevier Science.

Cheater, A. (ed.) (1999a) *The Anthropology of Power: Empowerment and Disempowerment in Changing Structures* (ASA Monograph 36) (pp. 1–12). London: Routledge.

Cheater, A. (1999b) Power in the postmodern era. In A. Cheater (ed.) *The Anthropology of Power: Empowerment and Disempowerment in Changing Structures* (ASA Monograph 36) (pp. 1–12). London: Routledge.

Church, A. and Coles, T. (eds) (2007a) *Tourism, Power and Space*. London: Routledge.

Church, A. and Coles, T. (2007b) Tourism and the many faces of power. In A. Church and T. Coles (eds) *Tourism, Power and Space* (pp. 269–284). London: Routledge.

Cohen, A. (1976) *Two-dimensional Man: An Essay on the Anthropology of Power and Symbolism in Complex Society*. Berkeley, CA: California University Press.

Coles, T. and Church, A. (2007) Tourism, politics and the forgotten entanglements of power. In A. Church and T. Coles (eds) *Tourism, Power and Space* (pp. 1–42). London: Routledge.

Colson, E. (1977) Power at large: Meditation on 'The Symposium on Power'. In R.D. Fogelson and R.N. Adams (eds) *The Anthropology of Power: Ethnographic*

Studies From Asia, Oceania, and the New World (pp. 375–386). New York: Academic Press.

Conklin, B.A. (1997) Body paint, feathers, and VCRs: Aesthetics and authenticity in Amazonian activism. *American Ethnologist* 24, 711–737.

Conklin, B.A. and Graham, L.R. (1995) The shifting middle ground: Amazonian Indians and eco-politics. *American Anthropologist* 97, 695–710.

Crick, M. (1994) *Resplendent Sites, Discordant Voices: Sri Lankans and International Tourism.* Chur: Harwood Academic Publishers.

De Kadt, E. (ed.) (1980) *Tourism: Passport to Development? Perspectives on the Social and Cultural Effects of Tourism in Developing Countries.* Oxford: Oxford University Press.

Elliot, J. (2004) *Politics of Tourism: A Comparative Perspective.* London: Routledge.

Eriksen, T.H. (2002) *Ethnicity and Nationalism: Anthropological Perspective* (2nd edn). London: Pluto.

Fogelson, R.D. and Adams, R.N. (eds) (1977) *The Anthropology of Power: Ethnographic Studies From Asia, Oceania, and the New World.* New York: Academic Press.

Franklin, A. (2003) *Tourism: An Introduction.* London: Sage.

Geertz, C. (1973) *The Interpretation of Cultures.* London: Fontana Press.

Gellner, E. (1983) *Nations and Nationalism.* Oxford: Blackwell.

Giddens, A. (1984) *The Constitution of Society.* Cambridge: Polity.

Gledhill, J. (1994) *Power and its Disguises: Anthropological Perspectives on Politics.* London: Pluto Press.

Hall, C.M. (1994) *Tourism and Politics: Policy, Power and Place.* Chichester: Wiley.

Hall, C.M. (2007) Tourism, governance and the (mis-)location of power. In A. Church and T. Coles (eds) *Tourism, Power and Space* (pp. 247–268). London: Routledge.

Hall, D. (ed.) (2004) *Tourism and Transition: Governance, Transformations and Development.* Oxford: CABI.

Hannam, K. (2002) Tourism and development I: Globalisation and power. *Progress in Development Studies* 2 (3), 227–234.

Harrison, D. (ed.) (2001) *Tourism and the Less Developed World.* Wallingford: CABI.

Ioannides, D., Apostolopoulos, Y. and Sonmez, S. (2001) *Mediterranean Islands and Sustainable Tourism Development: Practices, Management and Policies.* New York: Continuum.

James, W. (1999) Empowering ambiguities. In A. Cheater (ed.) *The Anthropology of Power: Empowerment and Disempowerment in Changing Structures* (ASA Monograph 36) (pp. 13–27). London: Routledge.

Kuper, A. and Kuper, J. (eds) (1985) *The Social Science Encyclopaedia.* London: Routledge and Kegan Paul.

Lea, J. (1988) *Tourism and Development in the Third World.* London: Routledge.

Lienhardt, R.G. (1966) *Social Anthropology.* Oxford: Oxford University Press.

Lukes, S. (2005) *Power: A Radical View* (2nd edn). Basingstoke: Palgrave Macmillan.

Mair, L. (1972) *An Introduction to Social Anthropology.* Oxford: Oxford University Press.

Mowforth, M. and Munt, I. (1998) *Tourism and Sustainability: New Tourism in the Third World.* London: Routledge.

Nash, D. (1989) Tourism as a form of Imperialism. In V.L. Smith (ed.) *Hosts and Guests: The Anthropology of Tourism* (2nd edn; pp. 37–54). Philadelphia, PA: University of Pennsylvania Press.

Nash, D. (1996) *The Anthropology of Tourism*. Oxford: Pergamon.

Richter, L.K. (1989) *The Politics of Tourism in Asia*. Honolulu, HI: The University of Hawaii Press.

Ritchie, J.R.B. and Crouch, G.I. (2003) *The Competitive Destination: A Sustainable Tourism Perspective*. Wallingford: CABI.

Robinson, M. and Boniface, P. (eds) (1999) *Tourism and Cultural Conflict*. Oxford: CABI.

Robinson, M., Evans, N., Long, P., Sharpley, R. and Swarbrooke, J. (eds) (2000) *Tourism and Heritage Relationships: Global, National and Local Perspectives*. Sunderland: Business Education Publishers.

Schultz, E. and Lavenda, R. (1995) *Cultural Anthropology: A Perspective on the Human Condition*. London: Mayfield.

Smith, V.L. (1989) *Hosts and Guests: The Anthropology of Tourism* (2nd edn). Philadelphia, PA: University of Pennsylvania Press.

Smith, V.L. and Eadington, W.R. (eds) (1992) *Tourism Alternatives: Potentials and Problems in the Development of Tourism*. Chichester: Wiley.

Telfer, D.J. and Sharpley, R. (2008) *Tourism and Development in the Developing World*. London: Routledge.

Tylor, E.B. (1871) *Primitive Society: Researches into the Development of Mythology, Philosophy, Religion, Language, Art and Culture*. London: John Murray.

Urry, J. (1990) *The Tourist Gaze*. London: Sage.

Wahal, S. and Pigram, J.J. (1997) *Tourism, Development and Growth: The Challenge of Sustainability*. London: Routledge.

Westwood, S. (2002) *Power and the Social*. London: Routledge.

Wilk, R. (1995) Learning to be local in Belize: Global systems of common difference. In D. Miller (ed.) *Worlds Apart: Modernity through the Prism of the Local* (pp. 110–133). London: Routledge.

Wolf, E. (2001) *Pathways of Power: Building an Anthropology of the Modern World*. Berkeley, CA: California University Press.

Wood, R. E. (1993) Tourism, culture and the sociology of development. In M. Hitchcock, V.T. King and M.J.G. Parnwell (eds) *Tourism in South-East Asia*. (48–70) London: Routledge.

Part 1

Tourism and the Power Struggle for Resources

DONALD V.L. MACLEOD

The struggle for resources is surely one of the ubiquitous features of human societies, and correspondingly, industrial endeavour. The tourism industry, in its multiple guises, relies upon a wide variety of resources, such as landscape, beaches and architecture, with which to attract visitors. It refers to such resources as products, in contrast to the visitors, which form the demand side of the equation.

Within this section of the book, the chapters examine different types of resources and the struggle for control over them, and it will become apparent to the reader that each case study has resonance for many tourism destinations worldwide. The resources examined are, in order of appearance: water, architecture, cultural heritage and the relatively intangible condition of security. The notion of security might also be understood as the condition that enables a destination to better utilise its assets, such as beaches, landscapes, material and intangible culture, and the indigenous population, in order to attract tourists: security will apply to all destinations as a desirable resource. In the chapter in question, terrorism threatens the security of the country, leaving it unable to capitalise on many other resources.

Power becomes involved where any struggle is experienced, as earlier discussions in the Introduction have explained. Different individuals or groups will have reasons for joining together or competing, be they economic, social, kinship related, religious, political, cultural or ideological. Tourism is a high profile and highly influential industry with immediate and profound impact on human groups. In this section, examples are given where disagreements and struggles occur between members of the same regional population who have varying relationships with and understandings of a resource. The chapters examine the following situations: people from widely different cultural backgrounds, cultural capital and economic status who ostensibly share a desire to conserve a resource; differing sectors of the nation who seek to promote their perspective on national and local history or culture; people from similar national and economic backgrounds who differ on religious and moral principles. In all these cases, tourism has acted as a complex catalyst or stimulant, often stirring up or emphasising existing contests and struggles for resources.

In Chapter 2, entitled 'Water Sports: A Tug of War over the River', Veronica Strang looks at the struggle for control over the use of water in Queensland, Australia: in particular, the different sub-cultural perspectives that farmers and recreational water users have towards water and its utilisation. Farmers wish to protect their allocation of water, which contrasts with tourists and the tourism industry's support for environmental groups who wish to see tighter regulation on the management of water. Strang points out how these two outlooks reflect deeply held values and ideological commitments, with economic rationalism being opposed by ecological environmentalism. Furthermore, the growth of urbanisation and urban populations has consequences for political power and domestic water consumption, outflanking the farming community. This is, moreover, bolstered by the increasing economic power of the tourism industry.

A detailed insight into the worldviews of the competing groups help the reader to comprehend the entrenched and widely differing attitudes towards water usage, which is shown to be exacerbated by the contemporary global post-industrial move from a production-based economy to one driven by consumerism and the service industry.

Strang draws our attention to the relationship between ownership and power and considers how tourism has led to more people experiencing a feeling of belonging in the landscape, which transforms into a sense of ownership – one that is collective and not formalised by legal title. In addition, this landscape and its water is perceived as directed towards the regeneration of self and biodiversity by the tourists and environmentalists, in contrast to its role in a farming landscape, which focuses on its ability to produce food and economic income.

In Chapter 3, entitled 'Heritage and Tourism: Contested Discourses in Djenné, a World Heritage Site in Mali', Charlotte Joy looks at the town of Djenné, which has been awarded the prestigious status based mainly on the architectural form of the vernacular housing and mosque and surrounding archaeology. This material culture is the attraction, the product at the heart of the development of tourism, which forms the mainstay of the economy for the town. The chapter reveals the link between tourist expectation and future strategies for the presentation of the town, which have remained largely undisclosed in relevant discourse. Joy gives a detailed description of the construction of the mud-brick houses and the crucial change in composition of the bricks with the consequential growth in the use of external tiling and cement, which are perceived by UNESCO as threatening the structure of the buildings and therefore the viability of the World Heritage Site status for the town.

The struggle for control over the construction and reparation of the houses is between the home owners who have economic and social reasons for using the tiles as opposed to the UNESCO representatives who wish to preserve traditional building patterns. Visitors, it is revealed, remain enchanted by the town. Part of the problem is the distinctiveness of the architecture, which becomes a reason for protecting it as the 'other', a position supported by UNESCO, an organisation perceived by some to be elitist and Eurocentric, and potentially symbolic of the power of the West.

Specific tourists search for authenticity in different ways, and Joy details a few that represent the individualistic types that are drawn to Djenné. Many tourists express sympathy with the plight of the villagers and become critical of the restrictions that may be imposed on building. At the same time, tourism has re-enfranchised some inhabitants of the town by enabling them to sell products, which were formally excluded by elitist readings of cultural heritage, a process partly helped by the tour-guides responding to the expectations and interests of the increasingly sophisticated tourists. As with Strang's example of control over water, control over material resources in Djenné is being competed for by two groups with differing ideologies, although in Djenné both groups are aware of the value of tourism.

Chapter 4 in this section is entitled 'Power, Culture and the Production of Heritage' and is written by myself, Donald Macleod. This chapter investigates the relationship between power, culture and heritage, examining these concepts, suggesting a model of power for approaching the subject and drawing on three different case studies based on fieldwork undertaken by the author. It explores the way groups in society represent cultural heritage and how they are able to promote aspects of power to their own advantage: patterns in such behaviour are noted and the contrast between official state-supported heritage and unofficial grassroots heritage is used as a way of analysing the material.

The first case study, La Gomera in the Canary Islands, looks at the way the Spanish colonial heritage, including architecture and the memory of Christopher Columbus, is maintained and promoted by state agencies, as well as the aspects of pre-Columbian culture, the Guanche people and their artefacts. This heritage is largely contained in the capital and port, San Sebastian, and is contrasted with the tourist destination area of Valle Gran Rey, where the memory of harsh times in the 1950s and the need to escape poverty is kept alive by a memorial to a large sailing boat, the 'Telemaco', which took some 171 people to the shores of Venezuela. There is also a newly erected statue of Hautecuperche, a Guanche rebel leader who assassinated an unpopular Spanish Count of La Gomera in the 15th century; this complements the mildly subversive memorial to Telemaco.

The second case study examines the Dominican Republic, again looking at the Spanish colonial heritage, this time in the capital Santo Domingo. It also comments on the political action that influenced the suppression of ethnic cultural history by omitting African heritage from a museum. The contrast between state-sponsored heritage items and the unofficial grassroots public manifestation is made clear by the example of the Brito family of Bayahibe village, who have created their own small memorial to 'The Founder of the Village' (a Brito ancestor), as well as the 'Protector' of the village (a politician). There is a strong irony in that Bayahibe is being promoted by tourism authorities as a 'fishing village' whilst tourism, in the form of motor-boat trips from the village port to a small island, is destroying the fishing economy.

The third and final case study of Dumfries and Galloway in Scotland, considers how the state agencies continue to promote Scottish icons, such as castles and Robert Burns, in the region; thus running the risk of leaving the region represented as a less distinctive part of Scotland than it actually deserves. Locally driven projects are given as examples, such as the product-based 'theme towns'. These may offer some hope of regeneration and help the region to strengthen its cultural heritage offerings and unique identity. However, the distinction between state-led and grassroots initiatives is blurred in this example because of the high involvement of agency-employed individuals in their working capacity and in their free time.

In summary, this chapter illustrates the inextricable relationship between power, culture and heritage and shows how the phenomenon of tourism is increasingly transforming the balance of power and opening opportunities to rework heritage resources for the advantage of various groups and individuals.

Finally, Chapter 5, entitled 'Cultural Perspectives on Tourism and Terrorism' by Michael Hitchcock and I Nyoman Darma Putra, considers security as a resource and examines how terror can be a form of

insecurity leading to extreme instability. The authors show that anthropologists are well placed to shed light on issues such as security and
terror, and this chapter illustrates the advantages of a deep understanding of the destination and its people, which long-term fieldwork
affords. It is argued that in order to develop and maintain a successful
tourism industry, political stability is a prerequisite. The cases of the Bali
bombings are particularly appropriate examples for exploring these
issues. The bulk of the research was done into the 2002 bombings and
reflects one of the most sustained investigations into attacks on tourists.
It emerges that the tourists themselves were targeted because they were
Westerners and non-Muslim, and especially perceived as American
allies. However, the authors, on examining the term Jihad suggest there
was a lack of authorisation given to the bombers, as the nation was not
itself under attack.

Security was highly prioritised by the government following the
bombings, especially in relation to visitors, and there was even a
dedicated programme of tourism training introduced for police officers.
However, the 2005 Bali bombings produced a different reaction from the
police, who were more secretive. Visitor numbers dropped drastically,
partly due to international media coverage. The authors conclude that
security needs to be developed as a resource and maintained. Local
knowledge is needed to explain the necessity for security and how to
deal with potential and actual attacks by terrorists, who have many
different motives as this chapter shows.

The sophisticated and detailed understanding of the sensitive subject
matter dealt with by Hitchcock and Darma Putra shows the importance
of ethnographic fieldwork and the drive to understand the worldview of
the local population, the 'emic' perspective. This has been part of the
anthropological project and forms an important component of the chapters
discussed above. It is by affording such insights into the indigenous
people, tourists, the host culture and the workers in the tourism industry,
that the anthropological research method of longitudinal fieldwork-based
ethnography demonstrates its immense value, as the following chapters in
this book will show. A rich, complex and subtle portrayal is given, an
interpretation whereby an understanding of the reality of concepts like
power, culture and heritage comes into a vision that details variety, and
enables the interested reader to get a better grasp of the situation.

Chapter 2

Water Sports: A Tug of War over the River

VERONICA STRANG

Introduction

As spiralling growth and development place unsustainable pressures on water resources around the globe, the right to control and manage river systems has become increasingly contested. Populations and urban areas are expanding fast, with more affluent lifestyles pushing *per capita* water use higher and higher. Farming activity is intensifying, and in every decade since the 1960s, countries such as the USA, Australia and New Zealand have doubled their use of water for irrigation. As Reisner (2001) points out, such expansions have been driven by an almost fanatical desire to 'green the desert' and 'civilise' putatively unproductive arid landscapes and make them productive in relation to human needs. The transformation of the desert in Israel is the most well-known example, but there are many other such endeavours around the world, for instance in the Middle East, America and Australia.

For a long time there was unquestioned acceptance that the irrigation of arid areas constituted a positive form of development. It was taken for granted that this should continue in the future with, for example, the USA looking to Canada to provide a diversion scheme from the Great Lakes, and Australia – even now – reviving earlier grandiose plans to turn major rivers inland to create new areas of farmland.[1] However, increasingly obvious environmental degradation is forcing a rethink about the wisdom of what has often proved to be only a short-term transformation, in which productive 'greening' has been rapidly followed by widespread land salination.[2] This is now a major problem in Australia, where vast swathes of land have been over-irrigated and are now so salinated as to be unable to regenerate even their original vegetation. Equally urgently, untrammelled development has forced industrialised societies to acknowledge that freshwater resources are finite. Built in the last half century to expand irrigation and to produce hydroelectric power, thousands of large dams have reduced many major rivers to sorry trickles. Thus, Reisner (2001: 5) notes the saline dribble of 'liquid death' that represents all that is left of the Colorado River as it crosses into Mexico and, in Australia, the Murray-Darling River basin offers a cautionary tale about the ecological costs of over-abstraction.

Accompanying the ecological problems caused by over-use of water resources are major social and cultural effects. Water shortages have led to increasingly bitter conflicts between upriver and downriver users, in particular where rivers cross national boundaries. Thus, Donahue and Johnston (1998) describe numerous transboundary battles over water in, for example, the Honduras, Canada, Zimbabwe, Japan and Israel. As they point out, competition for water can create major rifts within societies too:

> The story of water is all too often a story of conflict and struggle between the forces of self-interest and opportunities associated with "progress" and the community-based values and needs of traditional ways of life. (Donahue & Johnston, 1998: 3)

Implicit in this comment is the idea that there is an emerging pattern in which the demands of more powerful water-using groups override those of others, often dispossessing – and thus disenfranchising – minority indigenous or rural communities and destabilising long-standing modes of environmental engagement. To some extent, this reflects a fundamental tension between the expanding demands of 'the market' and efforts to maintain the more holistic processes of everyday social life. As Polanyi pointed out half a century ago:

> For a century the dynamics of modern society was governed by a double movement: the market expanded continuously but this movement was met by a countermovement checking the expansion in definite directions... This was more than the usual defensive behavior of a society faced with change; it was a reaction against the dislocation which attacked the fabric of society. (Polanyi, 1957: 130)

As Polanyi (1957: 178) made clear, the transformation of land and other resources into commodities was a vital part of this process. Contemporaneously, Wittfogel (1957) drew attention to the vital relationship between political power and the ownership or control of water, and this critical connection has been further elucidated historically and ethnographically in recent years (e.g. Worster, 1992; Blatter & Ingram, 2001; Cruz-Torres, 2004; Mosse, 2003; Strang, 2004; Swyngedouw, 2004). In broad terms, industrialisation and development, and the more recent globalisation of these activities, has resulted in a process of enclosure, commodification and privatisation of previously collective water and land resources by ever smaller power elites, and increasing competition for these resources.

Concurrently, in industrialised countries, there has been a steady process of urbanisation and a marked shift from manufacturing to service-based industries. The decline in farming in Australia mirrors similar reductions in its importance in the UK, Europe and North America. In essence, this is a move away from production to an emphasis on consumption, casting these societies primarily as the consumers of food, resources and products from poorer nations. Moving into a global food market – and thus away from self-sufficiency – has allowed affluent urban populations to reconsider their relationships with land and resources. One major consequence of this shift is a growth in recreational use of the environment, and a commensurate devaluation of agriculture and 'primary production'. In Australia, as elsewhere, the resultant rural-urban schism has been further widened in recent years by increasing demands for supplies from urban water users at the expense of farming and rural communities. Thus, in any tug of war over the river, recreational water users and farmers find themselves, with increasing frequency, on opposite banks.

These changes and tensions are amply demonstrated in Australia. Here, an intense ideological commitment to development is firmly rooted in colonial ideas about 'civilising' an 'untamed' landscape (and its people) through the application of a Christian work ethic and – as elsewhere – the transformation of an arid environment into suitably 'productive' farmland (see Carter, 1987; Schaffer, 1988; Strang, 1997). This developmentally orientated process has become more deeply entrenched in a contemporary political era dominated by economic rationalism and a largely unquestioned assumption that 'growth is good'. However, in the last few decades these driving forces have also brought water use and management in Australia to a state of crisis, creating major ecological problems and generating a range of tensions about who should control the rivers and their environs.

Numerous groups are empowered – or hope to be empowered – to direct water resources. Government departments (at a State and National level) have diverse and sometimes conflicting responsibilities in relation to water management: to protect the environment; to protect cultural heritage; to manage social issues; and to assist economic growth. Then there is a range of industries and sub-cultural groups anxious to gain or retain access to water and land: the agricultural and mining sectors;

environmental organisations; indigenous communities; and domestic water users. All are linked in complex networks of social, political and economic relationships. Within these networks, two of the most diametrically opposed groups are the farmers, for whom water alloca- tions are fundamental to their ability to maintain agency, identity and productivity; and recreational water users, whose very different engage- ments with water lead them to question the values of productivist agriculture and to support opposing aims. Their differences bring to the surface some of the key cultural and ideological issues that underlie many conflicts over land and resources.

This chapter, therefore, considers these two groups, their cultural engagements with water and their efforts to promulgate their own interests. It draws on long-term ethnographic research on two major river catchments in Queensland: the Brisbane River in South Queensland and the Mitchell River in Cape York.[3]

Two Rivers

Both the Mitchell and Brisbane Rivers support a range of economic and recreational activities. The Brisbane River flows southwards from the Jimna Ranges, which are mostly taken up with cattle properties, through several large dams that provide drinking water, hydroelectricity and irrigation for the south-east area of the State. Turning eastwards, it then passes through a lower valley that constitutes Queensland's most fertile farming area, winding finally into the extensive urban sprawl of the city, whose population makes extensive recreational use of the river and its environs. In the tropical north, the Mitchell River flows from the Great Dividing Range westwards to the Gulf of Carpentaria. A major irrigation scheme in the upper reaches of the river supports numerous small fruit and vegetable farms; then, in flatter savannah country, it flows through vast cattle stations, national parks and Aboriginal land, to culminate in a series of major wetlands on the western coast of the Cape. Being situated close to Cairns, the river catchment has recently become a key tourist area.

From the earliest days of European colonisation, the history of both catchment areas centred squarely on the establishment and development of pastoralism, agriculture and their related industries, and on other resource-based industries such as mining, quarrying and forestry (see Davie *et al.*, 1990; Gregory, 1996). The lengthy dominance of Joh Bjelke- Peterson's State Government had a major impact.[4] It ensured that, even more than other States in Australia, Queensland remained wedded to a highly positivist vision of growth and development in which productivist farming was valorised, environmental issues were given little consideration, and 'greenies' were regarded as dangerous cranks. In recent years, with a stronger presence of environmental issues in public discourses, rural producers have become more *au fait* with ecological concerns and increasingly willing to take these into account, but there remains a deep social and political chasm between them and the growing number of environmental groups.

Throughout Queensland's history, the economic centrality of primary producers meant that they were commensurately powerful in political terms, with rural landowners building on the early leadership of a colonial 'squattocracy' to take a dominant role in directing political events. Until relatively recently in Queensland, as in much of Australia, a significant number of Government and community leaders tended to be major landholders, and many of their aides also came from farming stock. Thus, in the political arena, and in contributing to a national identity and its representations, Queensland's voice was largely that of its rural communities.

This dominance remained unchallenged until the latter half of the 20th century, when post-war immigration initiated major growth in Australia's urban areas. With further influxes from other countries, these conurbations grew steadily over the next several decades. Echoing a widespread international trend, many people also moved from rural to urban areas as farming underwent a process of automation and amalgamation, which rapidly reduced the agricultural labour force. Urban expansion has been particularly evident in both north and south Queensland, boosted by a major population shift from New South Wales and Victoria. By the early 2000s, over a thousand people a week were either moving northwards to Brisbane in search of cheaper housing and new job opportunities, or retiring to the tropical areas of northern Queensland.

Both the Brisbane and Mitchell River catchments have therefore seen significant population growth. As a result, the urban vote – and thus the urban voice – has begun to speak much more loudly than that of the numerically diminished and economically shakier farming community. Critically, the domestic water needs of enlarging urban areas now represent major competition for limited water resources.[5] With several lengthy droughts in recent years, these demands have deprived farmers of all or part of their annual water allocations, placing them under intense economic and social pressure.

There have been other crucial changes in the last two decades. It has become inescapably apparent that over-abstraction in Australia's river catchments has caused extreme ecological problems (see Dennison & Abal, 1999). Concern about these and other kinds of environmental degradation[6] has encouraged the growth of an increasingly powerful environmental movement, led – as elsewhere – by an enlarging middle class (see Anderson & Berglund, 2002).

Urbanisation, in accord with the wider international trends noted previously, has also brought the emergence of numerous secondary industries, in particular tourism. Proliferating tourist developments (e.g. on Queensland's Gold Coast) have added considerably to urban demands for water resources – again taking precedence over farming allocations. And, as a substantial contributor to the State and National economy, the tourist industry is increasingly influential in the political arena and in decision making about environmental management. As Sheller and Urry (2004) observe, it has its own networks and relationships. To some extent, these operate separately – and sometimes oppositionally – to the longer established networks that support primary production.

There is an important relationship between the environmental movement and the tourist industry: indeed, it could be argued that, to some extent, they represent different facets of the same shift, by the majority of the population, away from direct forms of production, towards a more 'consuming' stance in which the material environment is engaged with aesthetically, spiritually, intellectually and recreationally. A globalising economy has allowed both environmentalists and tourists to interact with land and water in ways that, though by no means detached from economic realities, are framed as non-commercial activities. Most are employed in non-rural forms of production, inhabit urban areas and are free to consider and engage with the environment in largely ideological or recreational terms. Both groups (whose membership often overlaps) can afford to be critical of activities that are more focused on production, and which have impacts on the aspects of the environment that they valorise.

Cultural Landscapes and Engagements with Water

Each of the interest groups involved in contests for water have their own cultural land and waterscapes that map and evaluate the environment in particular ways. As Bender and others have made clear, cultural landscapes are fluid and dynamic, reflecting the beliefs, knowledge and values of the groups that create them (Bender, 1993; Hirsch & O'Hanlon, 1995; Strang, 1997). Some cultural landscapes provide more common ground than others, while major divergences lead to lively political contests (see Bender & Winer, 2001).

The cultural landscapes of the environmental movement and the tourist industry have some important commonalities and differences. Tourism in Australia, as elsewhere, readily encompasses built 'cultural heritage', valorising cityscapes (e.g. Sydney, the Harbour Bridge, the Opera House) and smaller local heritage sites, such as historic mine sites or pastoral homesteads, which are found in rural areas. However, it focuses much more heavily on what are regarded as 'natural' amenities: dramatic land and waterscapes; unique wildlife; large 'wilderness' areas. These are seen as the country's 'natural' assets (see Carrier & Macleod, 2005).

From the time when tourism began to emerge – first as a fairly low-key business – in the late 1970s and early 1980s, it combined attention to coastal 'sand and surf' activities with an interest in wildlife and 'nature'. The latter brought tourists inland, into areas previously dedicated fairly exclusively to mining, farming and pastoralism. Long ignored national parks (generally composed of areas that had been seen as having little use for other purposes) became camping destinations, and four-wheel drive expeditions into 'the outback' became increasingly popular. More recently, as in other tourist areas, there has been an enthusiastic shift towards 'ecotourism' and campsites and resorts focused on the appreciation of landscapes, birds and wildlife. This has brought the interests of the environmental movement and recreational land and water users into closer alignment.

Thus, the cultural landscapes created by the tourist industry and environmental organisations overlap considerably. Both cast as primary features areas in which there is a high concentration of wildlife and/or characteristics that are aesthetically pleasing. More often than not, such areas are seen as 'unspoilt' (i.e. undeveloped), thus framing any and all development in negative terms, and valorising only the 'natural' aspects of the environment. Given the varied plants and wildlife that they support, their 'scenic amenity' (Preston, 2001), the most highly valued places are those containing fresh water: rivers; creeks; wetlands; water-holes and so forth.

Recreational uses of water are highly diverse. On the domestic front, they often entail the sensory enhancement of familial space through the imaginative use of water to create cool green gardens and aesthetically pleasing water features (see Strang, 2004). As in many countries over the last century, people in Australia have also made increasing use of freshwater lakes and rivers for leisure activities: boating, swimming, fishing, jet-skiing and so forth (see Anderson & Tabb, 2002). Camping and social activities are heavily concentrated on riparian zones: for example in Brisbane itself, as in other towns along the Brisbane River, there are riverside boardwalks, bike tracks, parks and artificial beaches. In Cape York, the Mitchell River is central to the 'wilderness experience', offering peaceful campsites, exciting opportunities for fishing and hunting and other more benign opportunities to enjoy the local landscapes and wildlife. All of these 'amenities' are heavily used by local populations, as well as being intensively marketed to foreign tourists.

Despite their diversity, recreational engagements with water do have some important things in common. As noted above, although supportive of a large and profitable industry, they are not seen as directly commercial, in that the participants 'experience' the river, rather than directing its water resources into their own economic activities. All represent 'play' rather than 'work' and are thus centred upon sensory and aesthetic experiences that have been shown to engender affective relationships with places and greater concern for their ecological well-being (see Strang, 1996; Milton, 2002; Sheller & Urry, 2004). More pragmatically, close physical contact with (and often immersion in) water sources leads naturally to concerns about pollution from farms and industries and to related anxieties about over-abstraction and its impacts on water quality.

There are other aspects of recreational water use that generate support for environmental protection. As noted elsewhere, 're-creation' is also concerned with the renewal of the self and its identity (Strang, 1996). The terminology describing tourism reveals expectations that it will enable 're-vitalisation', 're-viving', 're-charging' and 're-storing', which centre on the idea that submergence in leisurely sensory experience reconstitutes the self and its life forces and is thus empowering. Therefore, at a superficial level, tourists express and affirm their identity and status through their ability to make luxurious, non-commercial use of the environment, and at a deeper, more embodied level, they engage with it experientially at a sensory and emotional level that is also 'productive' in that it re-produces the vitality of the self (see Csordas, 1994; Damasio, 1999). For some, this involves 'communing with nature' in a way that highlights aesthetic and spiritual ideas, and thus 'feeds the soul'. This is a powerful trope in ideas about wilderness, exemplified by Thoreau's writing, and focusing on the poetics of human environmental engagement. The close engagement with places that it permits encourages some degree of co-identification and emotional connection, encouraging a sense of belonging (see Abram *et al.*, 1997; Feld & Basso, 1996; Milton, 2002; Nast & Pile, 1998; Strang, 2005b). There are other ways to 'commune', as Reason says:

> Everywhere and always, human beings' discursive relation to the natural world is thematized through tropes and metaphors that govern the possibilities of intelligible representation of those relations. The natural environment serves as a screen on which we may project our fears, thrills, thralls, and ecstasies. Equally, the rhetorical resources of "relations with nature" provide for an idiom in which to represent oppositional identities. (Reason, 1998: 88)

McIntyre observes that there is an increasing trend towards 'adventure tourism' and a more challenging and adversarial interaction with 'nature':

> The general trend in visits to nature-based recreation areas appears to be one of increasing preference for more frequent trips of shorter duration... Current use of wilderness areas is marked by relatively high levels of activity, or participation in outdoor challenge programs... Embedded within the commodification and commercialization of the wilderness experience is the growing use of high technology devices in wilderness, such as hand-held GPS (Global Positioning System) and cellular phones, four-wheel-drive tour buses, mountain bikes, and the constantly expanding area of extreme sports... This analysis of current wilderness experiences indicates that for the majority of natural area visitors, the reality of the modern experience may be quite different from the one they are supposed to have – if the nature writers are to be believed. (McIntyre, 1998: 79)

Building on a well-established tradition of 'battling' against nature, this trend towards testosterone-laden 'bush-bashing' is very evident in Australia. McIntyre questions whether such cursory and commodified interactions can provide the emotional connection enabled by more 'spiritual' forms of engagement:

> The characteristics of modern society raise some doubts about the likely achievement of the values supposedly arising from wilderness experiences. Is it likely that today's wilderness user, cocooned in fibrepile and goretex, on a brief (1–2 day) trip into the wilderness, feels oneness, humility, and immersion? (McIntyre, 1998: 79)

The ethnographic evidence from Queensland, however, suggests that while such forms of activity may not induce 'humility', they still entail direct sensory engagement with places, and encourage a strong appreciation of their 'natural' qualities. At the very least, they imply a vested interest in access to unpolluted water and clean, green spaces. Thus, the recreational engagements with land and water conducted in the Brisbane and Mitchell River catchment areas share – to varying degrees – some important characteristics with those of environmental groups, engendering widespread support for their concerns.

An Agricultural Landscape

Water places are also central to an 'agri-cultural' landscape too, though obviously for different reasons. The exigencies of modern farming demand a pragmatic set of priorities. Access to and control of water are central to economic and social life in rural communities, and with the dominance of economic rationalism and involvement in a global market, rivers and other water sources have been increasingly quantified and commodified in accord with specifically economic aims and the potential 'use values' of resources.

The cultural landscapes of farmers and graziers, therefore, foreground water as irrigation sources and as water for stock, and river valleys are evaluated not as 'places to play' or 'places that generate wildlife', but as fertile, productive land. Soils are assessed according to whether they are water retaining or thin and friable. Land forms are considered as potential grazing pastures, crop fields and fodder meadows. Waterways are rated as reliable or sporadic, and valued according to their volume rather than their biota. Locales are therefore seen through a lens focused on systematic food production.

Nevertheless, although squeezed by the demands of the market into an increasingly narrow economic mode of engagement, farmers and graziers continue to have a much more complex relationship with land and water. Essential for settlement, cattle and crops, water places were the first places seized and inhabited by colonists and thus have the lengthiest historical associations for rural landowners.[7] In this sense, there is some common ground with other groups, in that many landowners have emotional connections to places that they see as integral to their long-standing social identity. Interviews with landowners also suggest that they often share tourists' and environmentalists' appreciation of land and waterscapes' aesthetic and 'natural' qualities, and of the wildlife that inhabits their land. However, farmers' social identity remains centred on production and a role as the 'food producers' on behalf of the wider Australian population. Their ability to generate wealth and health by using water to make food is fundamental to their sense of who they are, and it underlies strong values about the positive meanings of 'primary production' as a moral activity.

Historically, farmers felt secure in this role, and confident that their activities were valorised. Throughout the colonial era, they were the 'battlers' who tamed the arid land and made it support a new population; they were the 'heroes' who produced food for Australia's allies during the war; and they were the largest and most stalwart part of its post-war economy. The environmental movement's recent (and quite savage) environmental critique of their management of the land and water is received by many farmers as a painful betrayal that has undermined their sense of identity, while the accompanying regulation of their activities and competition for 'their' resources constitute a direct attack upon their power and agency. At the same time, the realities of a globalised economy make it difficult for them to resist the productivist values that have dominated the Australian political arena for many years. Until recently, the country was ruled by a right-wing National Government, politically sympathetic to these ideologies and very willing to deregulate industries accordingly. Although the Government changed in 2007, the market is still an overriding force, and farmers are under increasing pressure to continue to expand and intensify agricultural production – a pressure that is, of course, at odds with any aspirations to conserve ecosystems and reduce the over-use of water resources (see Carrier, 2004; Lawrence, 2004).

It is plain that with the development of such different ways of engaging with land and water, there has been some important diver-gence in the beliefs and values of the groups involved. Farming is under pressure to conform to the demands of economic rationalism, while greater recreational engagement has encouraged widespread support for the groups most critical of intensive productivist agriculture and its ecological effects. This broad political backing for ecological protection has also had a major effect on the priorities of Government, and the development of environmental policies. The result is an intensifying contest over who owns and controls land and water in Australia. In effect, in the use and management of Australia's rivers, the farmers are pulling in one direction – towards more irrigation and more production – while much larger groups are pulling in another direction, hoping to protect aquatic ecosystems and maintain an environment suited to recreational consumption.

A Tug of War

Farmers have a pretty clear vision of resource ownership and control. Most have either long-term leases or freehold title to land, and in many cases have done so for several generations. Though water ownership has proved somewhat less certain, they have managed to retain (and to some extent reaffirm) their rights to water (see Altman, 2004; Hussey & Dovers, 2006; Strang, in press). More fundamentally, they have a direct physical relationship with water, in which they abstract and channel it into irrigation technology and thus into the crops and livestock that represent their generative power. In this way, it is brought into their sphere of control and reconstituted as an extension of their individual and familial agency (Gell, 1998). It is thus ideationally as well as physically acculturated: transformed from being a commonly held 'natural' element into a privately owned and controlled 'cultural' commodity, which is central to their status and identity (see Goldman, 1998; Kopytoff, 1986). Therefore, it is unsurprising that this control should be passionately defended against any challenge, whether from environmentalists, recreational groups or other water users.

While farmers retained a positive role as the partners and agents of the larger population in managing the landscape and producing vital resources, their control of vital water resources was palatable. They could be seen as representing Australian society and as stewards of 'the commons'. Now, however, the social and economic conditions that separate rural and urban populations make such co-identification difficult: farmers and other 'primary producers' have become a distinct sub-cultural group in an increasingly multi-cultural, cosmopolitan and (above all) 'consuming' society. In this sense, they have been left behind and marginalised. With most food resources coming from elsewhere and a more adversarial social relationship between rural and urban communities, the issue of 'who controls the river' has become more contentious. For environmental groups and recreational water users, the farmers' desires to abstract and direct water from the commonly held river into their own 'self-interested' and increasingly privatised activities – at the expense of ecological health – has become far less acceptable.

Ownership is not merely a matter of title or material use: it can also be more broadly defined as having the power to affect decisions about what happens to people and things (see Strathern, 1999; Kalinoe & Leach, 2004; Verdery & Humphrey, 2004). Any more than a cursory form of engagement with place, with the investment of time and attention, tends to generate feelings of ownership. Recreational activities, especially those that involve regular visits, may be low-key in this regard, but they do have an effect. As McIntyre (1998) says, there is a strong trend towards greater frequency of recreational use of the environment. In Queensland, affluence has brought widespread access to four-wheel drive vehicles, and enabled the development of an ever more extensive network of sealed roads. Increasing numbers of city dwellers are therefore making regular forays into rural areas for recreational purposes. As well as building emotional connection, this greater involvement in the wider landscape undoubtedly encourages people to feel some degree of ownership of it and responsibility for its management, which challenges the long-standing ownership of the primary landholders.

This alternate 'ownership' is expressed in a variety of ways. Recreational land and waterscapes, while often presented as 'wilderness', are intensely humanised and extensively manipulated to fulfil tourist expectations. Land with prime views is cleared for camping, and roads are made to key sites. Supportive infrastructure and accommodation are put in place, and surroundings are manicured and enhanced. Places are mapped and named according to their particular characteristics: as heritage sites, wildlife parks, beauty spots, picnic places and so forth. All of these things shape and represent the environment in a way that foregrounds values associated with tourism rather than industrial farming, and all function to assert the primacy of a recreational cultural landscape.

More direct challenges over control emerge in squabbles about recreational access to land and water, and arguments about the extent to which 'leaseholders' have the right to exclude outsiders. Some parallels can be drawn with 'Right to Roam' battles in the UK and Europe, where the pressure of a large urban population on dwindling green spaces has led to extreme conflicts over access to land and waterways. Thus, many farmers and tourists in Queensland report angry exchanges over what can be framed either as access or as trespassing. For example, on the Palmer River (a tributary to the Mitchell) a notice assures tourists that a local grazier is not, in fact, allowed to wave a gun at them, and they should report any such incidents. A grazier further downriver recounts a joke about electrocuting 'trespassers' over his fence by means of a switch from his veranda (having erected a small sign saying that 'trespassing can be a shocking experience'), while an environmental activist describes the graziers themselves as 'trespassers' on the basis that 'they're just wrecking the country, completely wrecking it... they're trespassers. They shouldn't be there'.[8]

Different claims of ownership and the assertion of different values are similarly revealed in the priorities that people bring to debates. While farmers stress the need for material 'productivity', recreational water users regard the ability of river catchments to produce commercially profitable crops or livestock as secondary to their importance as land-scapes of putatively unspoilt 'nature' that support appealing wildlife and provide the opportunity for sensory and emotive connections with place. Thus, as Sarrinen (1998: 32) points out, they are inclined to prioritise 'non material values'. There is an important relationship between these non-material values and issues of ownership. 'Communing' with nature, in whatever mode, provides a form of symbolic social communion that hinges on ideas of oneness and harmony. There is a subtle but discernible link between this concept and ideas about communal ownership, most particularly of water. As observed elsewhere (Strang, 2004), water is the substance that links humans to each other and to their environments in a literal and metaphorical hydrological flow. Both symbolically and physically, it is intrinsically resistant to enclosure and exclusive forms of ownership, while at the same time standing as a powerful symbol of wealth and agency, and thus as an 'object of desire'. It can, therefore, be said to contain, in its very qualities, the basis for conflicting values between individual and collective interests. Achieving 'oneness' with place is connective, but it is also appropriative, laying claim, embedding identity and creating a sense of belonging that cuts both ways: 'I belong here' readily becomes 'here belongs to me'. Yet, because formal title is not part of this equation, the vision of ownership that emerges in a recreational landscape remains fundamentally collective in its form.

At the same time, the tourist industry itself suggests that this is somewhat a sleight of hand. Recreational water users want to regain 'free' and democratic access to collective resources, yet this is often enabled by a commercial industrial enterprise that is based on economic-ally 'rational' ideas about commodifying and selling aspects of the material environment. Focused on leisure and pleasure, tourism epito-mises consumption. Indeed, it is the apogee of affluence: both a reward for successful commercial enterprise and a successful commercial enterprise in itself. In this sense, it could be said to share conceptual space with agriculture, merely acculturating and 'producing' a different set of objects and creating direct commercial and political competition for the ownership and control of land and resources. It has been sufficiently successful in this regard to gain considerable power and influence, and to have some effect on decisions about where water resources are directed, thus major tourist enterprises in Queensland's south-east are able to gain priority over the needs of fruit and vegetable farmers upriver.[9]

Governance and Management

As noted in the introduction to this volume, political and economic processes reflect cultural change. The changing political economy of Queensland, and the shifting fortunes of tourism and farming, are readily apparent in recent developments in the way that water is governed and managed. Wilson (2004) describes a move towards 'post-productivist' governance. Over the last two decades, in formal terms, there has been an exponential increase in the regulation of water and how it is abstracted or impounded. Where farmers were previously free to drill water bores or place pumps in water courses, such abstraction must now be licensed. Meters are being placed on abstraction technology, along with volumetric charges for water. The Environmental Protection Agency has gained power, while the Department of Primary Industries has declined in influence, and wide-ranging environmental legislation has been introduced to protect aquatic ecosystems, with stringent regulations to prevent fertiliser run-off and other forms of water pollution, land degradation and the over-abstraction of water from the upper reaches of rivers.

This shift in political power and control over resources is manifested at a local level with a rapid proliferation of 'catchment management groups' of various kinds, in general aiming to tackle the ecological problems created by agricultural and other industrial activities within river catchments, and to ameliorate the effects of urban expansion. Brisbane City itself has over 30 local catchment groups, composed of people living alongside the many tributaries to the river. The Mitchell River had one of the first such organisations, the Mitchell River Watershed Catchment Management Group (MRWCMG). This was established in the early 1990s by the Aboriginal community at the estuary of the river, because of their concerns about the impacts of other activities, including recreational fishing, in the watershed (see Strang, 1997, 2001). The MRWCMG has since been carried forward by a range of 'stakeholders' including those committed to advancing tourism in northern Queensland.

Reflecting a larger contest for the control of water, between the State and Federal Governments, local catchment groups have recently been subsumed by the creation of larger regional catchment groups, set up and funded by the Federal Government to empower 'local stakeholders'.

Both at a local and regional level, catchment groups are generally dominated by people whose training is in environmental management, and whose major interest and expertise is in the health of Australia's aquatic ecosystems. They often work closely with the increasingly powerful Government departments that have a remit to manage and protect natural resources.

Primary producers have resisted this shift in managerial influence in a variety of ways: participating in rural Land Care associations to re-establish their own leadership in environmental care; forming their own Water Users groups to lobby on their behalf, and often joining local catchment groups to ensure that their own interests as 'stakeholders' are not overridden. These efforts represent the upholding of particular sub-cultural ideas and values that are increasingly divergent from those of the larger urban population and, most particularly, those encouraged by recreational uses of the rivers and their surrounding landscapes.

Opposing Engagements

Despite their apparent polarisation, both farming and recreational engagements with land and water are riven by conflicting social and economic pressures and divergent values and ideologies. Both contain tensions between production and consumption, between affective, complex relations with place and economically rational perspectives, and between traditional social values and more cosmopolitan life-styles. The two groups differ, however, in the weight that they assign to these issues, and these different emphases place them on divergent cultural trajectories in their activities, in their discourses and in their representations.

Water is encoded with powerful cross-cultural themes of meaning as a generative substance that is vital to the creation of wealth and health, and that enables social, spiritual, cultural and material reproduction.[10] For the farmers, these meanings are readily located within a productivist vision of agriculture in which water is directed to fulfil their aims by generating a variety of material resources. They are historically and materially committed to production, and heavily constrained by a market that demands the privatisation and enclosure of land and resources. This trajectory places them in conflict with their own, more complex, relationships to place and community, and with the larger urban population on which they depend.

For tourists, recreational water users and environmentalists, the generative power of water is directed towards self-regeneration and the generation of biodiversity. In consuming these, recreational water users are enabled by commercial enterprises and participation in a global economy to construct affective and putatively non-commercial relations with places and aspire to a more collective vision of human environmental engagement. Thus, they arrive at very different ideas about what rivers and water resources are for, and how these should be used and managed, and are willing to lend support to environmental policies and post-productivist governance.

The result is an intensifying political competition for power and agency in relation to land and, most particularly, water. Moreover, it seems likely that, as the need for local food production is further eroded by global markets, and recreational use of the environment continues to gain ascendance, rural agricultural communities and recreational water users will find their opposing aims increasingly difficult to reconcile.

Notes

1. For example, the Fitzroy River in northern Australia is regarded (by a small but vocal lobby) as a prime candidate for this kind of engineering.
2. Salination of land is a complex process, but often occurs when trees (whose deeper roots keep saline water below the soil surface) are removed and replaced with shallower-rooted irrigated crops. The problem is also caused by saline groundwaters being drawn to the surface for irrigation purposes, which spreads salts upon the soil and reduces the amount of water being filtered below the surface to recharge aquatic systems.
3. I have worked in the Mitchell River catchment for many years (see Strang, 1997), therefore this research builds on previous fieldwork. More recently, both river catchments were field sites in a three-year project funded by the Australian Research Council. This involved extensive ethnographic research on water issues in both areas. Over 300 people (a cross-section of water users) were interviewed during the course of the project, some of them several times, and a range of other standard anthropological methods of data collection and analysis were employed.
4. Bjelke-Peterson held office as the Premier of Queensland from 1968 to 1987.
5. Domestic water needs and water for power generation take formal precedence over farming allocations. Thus, when there is insufficient water, farmers do not get the annual allocations to which they are entitled, even though they may still be paying a flat rate to retain this entitlement.
6. For example, soil erosion; the pollution of aquatic systems; feral animals and plants; deforestation.
7. As noted elsewhere, they also have primary importance in the cultural landscapes of indigenous Australians (Morphy, 1993; Strang, 1997).
8. These examples are drawn from recent fieldwork in the area. Although none of these informants requested anonymity, it is perhaps best not to provide their names.

9. Similar dynamics are evident in other parts of the world: for example, in California, golf courses are now successfully competing with citrus farmers for water, and in Mexico, new hotels are depriving campesinos of scarce water resources.
10. See Douglas (1973), Bachelard (1983), Lakoff and Johnson (1980), Illich (1986), Rothenberg and Ulvaeus (2001) and Strang (2004, 2005a, 2006).

References

Abram, S., Waldren, J. and Macleod, D. (1997) *Tourists and Tourism: Identifying with People and Places*. Oxford, New York: Berg.

Altman, J. (2004) Indigenous interests and water property rights. *Dialogue* 23 (3), 29–34.

Anderson, D. and Berglund, E. (eds) (2002) *Ethnographies of Conservation: Environmentalism and the Distribution of Privilege*. New York: Berghahn Books.

Anderson, S. and Tabb, B. (eds) (2002) *Water, Leisure and Culture: European Historical Perspectives*. Oxford, New York: Berg.

Bachelard, G. (1983) *Water and Dreams: An Essay on the Imagination of Matter* (E. Farrell, trans.). Dallas, TX: Pegasus Foundation.

Bender, B. (ed.) (1993) *Landscape: Politics and Perspectives*. Oxford: Berg.

Bender, B. and Winer, M. (eds) (2001) *Contested Landscapes: Movement, Exile and Place*. Oxford, New York: Berg.

Blatter, J. and Ingram, H. (eds) (2001) *Reflections on Water: New Approaches to Transboundary Conflicts and Cooperation*. Cambridge, MA, London: MIT Press.

Carrier, J. (2004) *Confronting Environments: Local Understanding in a Globalizing World*. Walnut Creek, CA: Altamira Press.

Carrier, J. and Macleod, D. (2005) Bursting the bubble: The socio-cultural context of ecotourism. *Journal of the Royal Anthropological Institute* 11 (2), 315–334.

Carter, P. (1987) *The Road to Botany Bay: An Essay in Spatial History*. London, Boston, MA: Faber and Faber.

Cruz-Torres, M. (2004) *Lives of Dust and Water: An Anthropology of Change and Resistance in Northwestern Mexico*. Tucson, AZ: University of Arizona Press.

Csordas, T. (ed.) (1994) *Embodiment and Experience: The Existential Ground of Culture and Self*. Cambridge: Cambridge University Press.

Damasio, A. (1999) *The Feeling of what Happens: Body and Emotion in the Making of Consciousness*. New York: Harcourt Brace.

Davie, P., Stock, E. and Choy, D. (eds) (1990) *The Brisbane River: A Source-book for the Future*. Brisbane: Australian Littoral Society Inc. in association with The Queensland Museum.

Dennison, W. and Abal, E. (1999) *Moreton Bay Study: A Scientific Basis for the Health Waterways Campaign*. Brisbane: South East Queensland Regional Water Quality Management Strategy.

Donahue, J. and Johnston, B. (eds) (1998) *Water, Culture and Power: Local Struggles in a Global Context*. Washington, DC: Island Press.

Douglas, M. (1973) *Natural Symbols: Explorations in Cosmology*. London: Barrie and Jenkins.

Feld, S. and Basso, K. (eds) (1996) *Senses of Place*. Santa Fe, NM: School of American Research Press.

Gell, A. (1998) *Art and Agency: An Anthropological Theory.* Oxford: Clarendon Press.

Goldman, M. (ed.) (1998) *Privatizing Nature: Political Struggles for Global Commons.* London: Pluto Press.

Gregory, H. (1996) *The Brisbane River Story: Meanders Through Time.* Brisbane: Australian Marine Conservation Society.

Hirsch, E. and O'Hanlon, M. (eds) (1995) *The Anthropology of Landscape: Perspectives on Place and Space.* Oxford: Clarendon Press.

Hussey, K. and Dovers, S. (2006) Trajectories in Australian water policy. *Journal of Contemporary Water Research and Education* 135, 36–50.

Illich, I. (1986) *H_2O and the Waters of Forgetfulness.* London, New York: Marion Boyars.

Kalinoe, L. and Leach, J. (eds) (2004) *Rationales of Ownership: Transactions and Claims to Ownership in Contemporary Papua New Guinea.* Wantage: Sean Kingston.

Kopytoff, I. (1986) The cultural biography of things: Commoditization as process. In A. Appadurai (ed.) *The Social Life of Things: Commodities in Cultural Perspective* (pp. 64–91). Cambridge: Cambridge University Press.

Lakoff, G. and Johnson, M. (1980) *Metaphors We Live By.* Chicago, IL, London: University of Chicago Press.

Lawrence, G. (2004) Promoting sustainable development: The question of governance. Paper to XI World Congress of Rural Sociology, Trondheim, Norway.

McIntyre, N. (1998) Person and environment transactions during brief wilderness trips: An exploration. In United States Department of Agriculture, *Personal, Societal, and Ecological Values of Wilderness: Congress and Proceedings on Research, Management, and Allocation* (Vol. 1; pp. 79–83). Forest Service Proceedings. Fort Collins, CO: Rocky Mountain Research Station, RMRS-P-4.

Milton, K. (2002) *Loving Nature: Towards an Ecology of Emotion.* London, New York: Routledge.

Morphy, H. (1993) Colonialism, history and the construction of place: The politics of landscape in Northern Australia. In B. Bender (ed.) *Landscape, Politics and Perspectives* (pp. 205–243). Oxford, New York: Berg.

Mosse, D. (2003) *The Rule of Water: Statecraft, Ecology, and Collective Action in South India.* Oxford: Oxford University Press.

Nast, H. and Pile, S. (eds) (1998) *Places Through the Body.* London, New York: Routledge.

Polanyi, K. (1957) *The Great Transformation.* Boston, MA: Beacon Press.

Preston, R. (2001) *Scenic Amenity: Measuring Community Appreciation of Landscape Aesthetics at Moggill and Glen Rock.* Brisbane: Queensland Government.

Reason, D. (1998) Reflections of wilderness and pike lake pond. In United States Department of Agriculture, *Personal, Societal, and Ecological Values of Wilderness: Congress and Proceedings on Research, Management, and Allocation* (Vol. 1; pp. 85–89). Forest Service Proceedings, Fort Collins, CO: Rocky Mountain Research Station, RMRS-P-4.

Reisner, M. (2001) [1986] *Cadillac Desert: The American West and its Disappearing Water.* London: Pimlico.

Rothenberg, D. and Ulvaeus, M. (eds) (2001) *Writing on Water.* Cambridge, MA, London: MIT Press.

Sarrinen, J. (1998) Wilderness, tourism development and sustainability: Wilderness attitudes and place ethics. In United States Department of Agriculture, *Personal, Societal, and Ecological Values of Wilderness: Congress and Proceedings on Research, Management, and Allocation* (Vol. 1; pp. 29–33). Forest Service Proceedings, Fort Collins, CO: Rocky Mountain Research Station, RMRS-P-4.

Schaffer, K. (1988) *Women and the Bush: Forces of Desire in the Australian Cultural Tradition.* Cambridge, Melbourne: Cambridge University Press.

Sheller, M. and Urry, J. (eds) (2004) *Tourism Mobilities: Places to Play, Places in Play.* London: Routledge.

Strang, V. (1996) Sustaining tourism in Far North Queensland. In M. Price (ed.) *People and Tourism in Fragile Environments* (pp. 51–67). London: John Wiley.

Strang, V. (1997) *Uncommon Ground: Cultural Landscapes and Environmental Values.* Oxford, New York: Berg.

Strang, V. (2001) Negotiating the river: Cultural tributaries in Far North Queensland. In B. Bender and M. Winer (eds) *Contested Landscapes: Movement, Exile and Place* (pp. 69–86). Oxford, New York: Berg.

Strang, V. (2004) *The Meaning of Water.* Oxford, New York: Berg.

Strang, V. (2005a) Common senses: Water, sensory experience and the generation of meaning. *Journal of Material Culture* 10 (1), 93–121.

Strang, V. (2005b) Knowing me, knowing you: Aboriginal and Euro-Australian concepts of nature as self and other. *Worldviews* 9 (1), 25–56.

Strang, V. (2006) Aqua culture: The flow of cultural meanings in water. In M. Leybourne and A. Gaynor (eds) *Water: Histories, Cultures, Ecologies* (pp. 68–80). Nedlands: University of Western Australia Press.

Strang, V. (2009) *Gardening the World: Agency, Identity, and the Ownership of Water.* Oxford, New York: Berghahn.

Strathern, M. (1999) *Property, Substance and Effect: Anthropological Essays on Persons and Things.* London: Athlone Press.

Swyngedouw, E. (2004) *Social Power and the Urbanization of Water: Flows of Power.* New York: Oxford University Press.

Thoreau, H.D. (1971) *Walden.* Princeton, NJ: Princeton University Press.

Verdery, K. and Humphrey, C. (eds) (2004) *Property in Question: Value Transformation in the Global Economy.* Oxford, New York: Berg.

Wilson, G. (2004) The Australian landcare movement: Towards 'post-productivist' rural governance. *Journal of Rural Science* 20, 461–484.

Wittfogel, K. (1957) *Oriental Despotism: A Comparative Study of Total Power.* New Haven, CT: Yale University Press.

Worster, D. (1992) *Rivers of Empire: Water, Aridity and the Growth of the American West.* Oxford, New York: Oxford University Press.

Chapter 3
Heritage and Tourism: Contested Discourses in Djenné, a World Heritage Site in Mali

CHARLOTTE JOY

Introduction

This chapter explores the effect tourism has on the conceptualisation and management of cultural heritage in the town of Djenné, a World Heritage Site in Mali. It is based on ten months fieldwork in Djenné between 2004 and 2006 and a two-month internship in the Intangible Heritage Department of UNESCO in Paris in 2005. Since the town's inclusion on UNESCO's (United Nations Educational, Scientific and Cultural Organisation) World Heritage List, a debate has arisen about what its World Heritage status means, and the consequences for the lives of ordinary residents (the Djennenkés). Tourism is the economic engine of the town, yet much of the discourse around heritage preservation does not explicitly acknowledge the link between tourist expectations and future strategies to preserve the town's cultural heritage. Additionally, tourist expectations are understood within an archival apprehension of Djenné as a monument, instead of the more dynamic reality of tourism in the town.

In 1988, the Old towns of Djenné, consisting of approximately 1850 mud brick houses (Bedaux et al., 2003: 53) and its mud mosque, together with a 4-km radius of surrounding archaeological sites were judged as meeting two criteria necessary for inscription on UNESCO's World Heritage List: criteria (iii) to bear a unique or at least exceptional testimony to a cultural tradition or to a civilization which is living or has disappeared, and criteria (iv) to be an outstanding example of a type of building, architectural or technological ensemble or landscape which illustrates (a) significant stage(s) in human history.

The two elements of cultural heritage present in Djenné of special note for UNESCO are, therefore, its architecture and its archaeology. Since its inclusion on the World Heritage List, Djenné's architecture has been evolving and changing. Changing architectural practices, especially the use of fired clay tiles described below, are having a transformative effect on the outward appearance, the façade, of the town. As will be described, it is this very façade that has long been the focus of the Western gaze and fascination with Djenné. Consequently, it is assumed that tourists coming to Djenné do so primarily to see Djenné's world famous architecture. However, I will argue that many tourists coming to Djenné are motivated by the desire for an emotional authenticity that takes them beyond a purely visual apprehension of the town.

Djenné's Architecture

As well as the monumental mud Mosque, Djenné's houses are typified by three styles of façades: la Façade Marocaine, la Façade Toucouleur and the undecorated façade. The undecorated façade may be so by design, or may have become so through lack of upkeep (Maas & Mommersteeg, 1992). The Façade Toucouleur is differentiated from the Façade Marocaine through the overhanging 'screen' on the front door. The element of continuity in Djenné's architecture is the barey-ton, the mason association, who build and maintain all the houses in Djenné (Marchand *et al.*, 2003). One of the most notable changes in architectural practice in the last few hundred years has been the technology used by the masons to build mud bricks: from the pre-colonial Djenné-Ferey to the current use of Toubabou-Ferey. A second fundamental change has been the choice of materials used for making bricks.

Before colonial times, the masons of Djenné used hand-moulded cylindrical bricks called Djenné-Ferey. A retired mason in Djenné reported that the old bricks were ideally made from a mixture of mud, rice husks, beurre de karité (shea butter), a powder made from the fruit of the néré tree and a powder made from the fruit of the baobab tree. This mixture would be broken down with the use of animal urine and dung until it was ready to be moulded and 'baked' in the sun. These bricks are remembered by today's masons as more hardwearing than contemporary Toubabou Ferey bricks. Toubabou Ferey are square bricks, made using wooden moulds and, due to their shape, easier to manufacture and build with. They are usually only made of mud and rice stalks, as the other ingredients are now too expensive, such as the beurre de karité that has a high commercial value as the base for beauty products and other manufactured goods.

A fundamental change in agricultural practice that has led to the degradation of the quality of the bricks used in Djenné has been the introduction of mechanical rice de-huskers. These machines, which can be seen working at the river's edge, reduce the rice husks to powder, a far less appropriate ingredient for bricks than the manually extracted husk. The husks are used to bind the mud in a way that the powder does not achieve. Masons are reduced to importing at great expense, rice husks from outside Djenné and substituting, in part or in whole, the stalks of the plants for the rice husks. Additionally, the exposed stalks prove to be irresistible meals for the animals living in the compound, so some accessible parts of peoples' homes are simply nibbled away. Today, the only occasions when houses in Djenné are built with traditional ingredients, such as the beurre de karité, are when they are commissioned by the handful of rich expatriates in the town (see Plate 3.1).

Plate 3.1 Traditional house in Djenné

Building to such high standards is beyond the means of the ordinary Djennenké.

Decreased rainfall and river levels and the subsequent impact on fish stocks has led to a weaker concentration of calcified fish bones that, reputedly, made the mud resistant. Additionally, pollution in the river in the form of strands of plastic bags and bottle tops are now found in the mud and simply applied to the façades of the houses during the annual protective re-layering of mud. The houses in Djenné require an annual or biennial re-layering of mud (crépissage) to protect them from the rain. Houses that are not regularly maintained will rapidly fall into ruin.

These factors have led to the increasingly common practice of using fired clay tiles to protect the outside of Djenné houses. The tiles are fixed to the house using cement, a practice that many heritage experts warn can fundamentally undermine the entire structure of the building. UNESCO, through the work of the Cultural Mission in Djenné, vocally opposes the use of tiles, as it threatens Djenné's World Heritage Status, yet the practice is gaining popularity amongst residents because it promises to solve the regular problem of finding upkeep funds. Despite the initial high outlay, tiled houses can last for years without needing maintenance, meaning peace of mind for its occupants. The difficulty with accumulating wealth in Djenné is another reason for people tiling their homes when they have the available funds.

In the past, the use of tiles was reserved for the rich people of Djenné or only for use on the floors of houses. Now, even less well-off residents tile the outside of their houses because they feel that in the long term it makes more financial sense to do so. However, a study undertaken by a French architect in conjunction with Djenné Patrimoine in 2006 revealed that tiling houses was 14 times more expensive than using banco with added beurre de karité, and 18 times more expensive than using ordinary banco (2800 CFA/£2.80 per square metre for tiles as opposed to 196 CFA/£0.19 for banco with beurre de karité and 152 CFA/£0.15 for simple banco).[1] Therefore, the evidence suggests that despite an intuitive feeling amongst some Djennenkés that the use of tiles is a good long-term investment, the reality may be very different.

A controversial dimension to the debate surrounding the upkeep of houses is the fact that local authorities in the area, operating with restricted budgets, cannot meet the costs of upkeep on mud buildings and, therefore, build their administrative buildings, on the outskirts of town, in cement. Similarly, administrative buildings within the town are undermining a conservation message. In one instance, the house of the Sous-Préfet near the Campement Hotel in Djenné was covered in tiles, leading to the intervention of the Cultural Mission, which insisted on their removal. As a compromise, the Cultural Mission has agreed to pay for the ongoing upkeep of the house.

Amongst Djenné residents, there is a high level of awareness of the attraction of the town's architecture to tourists. The masons themselves are somewhat resentful that they do not derive more direct benefit from tourism for their labour, and in the past have demanded a share of the tourist tax (paid by each tourist entering the town) to better reflect the importance of their role within the tourist equation. As one mason told me:

> That's why tourists come. To see the houses, built one or two hundred years ago, in banco, still standing. Well, that is what fascinates (people), that is what attracts them. It's what gives us our value.

With the increasing cost of living in the town, and new technologies (such as the mechanical rice de-huskers), all parts of the population are updating and changing their subsistence practices. For example, while in the past, fishermen used string nets costing 50,000 CFA (£50 for 100 m of net), the majority now opt for a cheaper plastic version costing only 15,000 CFA (£15). Not only is the plastic version cheaper, but it is also more efficient at catching fish. However, unlike changes in architectural practices, the changes in fishing practices go uncommented upon by heritage officials in the town. A frame of reference for what can and cannot change within the town is therefore inconsistent and sometimes contradictory.

Djenné's architecture is problematic. On the outside, its façade is a source of national pride and inspiration, while on the inside, people often face challenging living conditions. These poor living conditions are present in some of the early writings about the town. In his descriptions of a traditional house in Djenné, Charles Monteil (1903), a French colonial administrator, follows his description of the façades of the houses with the qualification:

> Such is the general style of these houses which, from the outside, appear luxurious; but in reality are not comfortable due mainly to the lack of ventilation in the rooms... The darkness and dirt mean that these houses are the refuge of choice for vermin, insects, scorpions, rats and other animals whose closeness, while always disagreeable, is sometimes dangerous. (Monteil, 1903: 135; my translation)

Monteil's reaction to his direct experience of the town can be contrasted with an 'archival' apprehension of Djenné. The interest in the region's vernacular architecture has grown from the time of the arrival in the West of early photographs, postcards (Gardi, 1994) and colonial exhibitions (Morton, 2000). These glimpses gave the impression of Djenné as a faraway place, difficult to get to and to get around, with winding streets and compact architecture. In 1931, L'Exposition Coloniale Internationale, held in Vincennes in Paris, featured amongst its prized exhibits a 'Rue de Djenné'. Morton (2000) explains that the reproduced Djenné Mosque and streets were intended to stand for a primitive stage of architectural sophistication and were juxtaposed with architecture found in other parts of the French Empire, such as Indochina. The real-life poverty and French colonial experience in West Africa were simply airbrushed out of the picture depicted for the audience. The exaggeration of certain aspects of the Mosque, such as the wooden beams that were made to strut out decoratively, was a conscious attempt on behalf of the architects to convey the 'exoticism' of West African architecture.

Today, Djenné is still enchanting because it promises to be so different from the places people visiting it have come from (Hudgens & Trillo, 1999). This form of apprehending Djenné as other, or distant, is principally about fear and loss and dominated by a desire to record and protect. It can be found at the heart of UNESCO's project in Djenné.

UNESCO and Djenné

UNESCO's critics accuse the organisation of operating within a eurocentric and elitist conception of cultural heritage (Dutt, 1995; Eriksen, 2001; Fontein, 2000; Olaniyan, 2003; Singh, 1998; Turtinen, 2000). In part, this is due to large sections of the world being underrepresented within UNESCO's map of World Heritage, but it is also provoked by UNESCO's archival approach to the protection of cultural heritage. Describing UNESCO's work in Africa, Olanyian states:

> Employing instruments both of persuasion and coercion such as funding and legal statutes, the goal of such efforts is to identify, catalogue and conserve such monuments in their original locations as part of a vast network of a decentralized global museum. (Olanyian, 2003: 28)

Following UNESCO's adoption in 1972 of the 'Convention Concerning the Protection of the World's Cultural and Natural Heritage' (widely known as the World Heritage Convention), a 'World Heritage List' was begun in 1978 in order to identify and protect sites of 'outstanding universal value' throughout the world. As of July 2007, there are 851 properties inscribed on the list, of which 660 are cultural, 166 are natural and 25 are mixed sites.[2] Only 71 of these sites can be found in sub-Saharan Africa, leading to the launch in 2006 of an 'Africa Fund' to help African State Parties prepare national inventories and nomination dossiers (known as candidature files).

UNESCO relies on outside experts, such as ICOMOS (International Council on Monuments and Sites), to assess candidature files. These experts, in turn, rely on a dossier put together by a selected (usually elite) part of the population of the member state putting forward their submission. The candidature file follows a fairly strict template approach and is illustrated by carefully selected photographs. Additionally, the nomination and selection of World Heritage Sites within UNESCO is a political act, carefully balancing the needs of the organisation (UNESCO and its mandate) and those of the member states.

In order to fulfil its obligations to UNESCO, the Malian government established the Cultural Mission in Djenné in 1994. Cultural Missions were also established in the country's two other World Heritage Sites at the time, Bandiagara (Dogon Country) and Timbuktu.[3] At first, as their names indicate, the Cultural Missions were supposed to be temporary services put in place to help local populations best manage their own cultural heritage. Now, the Cultural Missions have become permanent structures, partly due to their success, but also due to the recognition that their 'mission' has not yet been accomplished. UNESCO helps the personnel of the Cultural Mission indirectly, principally through training programmes.

Contrary to some peoples' beliefs in the town, UNESCO does not give money to the Cultural Mission directly. Indirectly, however, money is channelled through UNESCO to undertake projects such as the 'Plan de gestion de la ville' (management plan). UNESCO also facilitates the Cultural Mission's work with outside partners by providing access to the Centre de Patrimoine Mondiale, ICCROM (International Centre for the Study of the Preservation and Restoration of Cultural Property) and ICOMOS for specialist knowledge, and the Getty Foundation, the Dutch Government and the Aga Khan Trust, who can all be counted on to finance projects. UNESCO also has a countywide presence through an office in Bamako, where they regularly organise conferences, usually for training purposes.

In terms of local politics, the Cultural Mission, by its own admission, is often disliked by the local population as it has to impose legislation that comes from the outside. This legislation states that no material changes may be made to Djenné's architecture without undermining the architectural integrity of Djenné as a whole, the town as a monument. Tourists coming to Djenné are not usually aware of the background to the complexities of the 'UNESCO debate' going on in the town, however, as I will argue, many of them very quickly attain an 'emotional' understanding of the town, which fundamentally brings into question whether seeing Djenné as a 'monument' is the motivation behind their journey.

The Tourist

An exploration of the motivations of tourists who visit Djenné allows a conceptual link to be made between UNESCO's pre-occupation with the preservation of the town and the demands of the tourist industry. It is assumed that tourists come to Djenné to see its unique architecture and archaeology. While this may be the case for some, many combine this with a search for an emotional authentic experience, a search that is often unsettled by the reality of the town.

Using Urry's (2002: 94) definition of the three important categories defining tourists sites, Djenné can be characterised as historical, romantic and authentic. Urry's analysis is centred on a Foucauldian concept of an organised and systematised 'tourist gaze'. Despite the limitations of the concept (Urry, 2002: 145), its application to Djenné is particularly illuminating. Although through UNESCO's work Djenné has been described as very much the object of the collective (and international) gaze, a tourist travelling to Djenné tends to be seeking the romantic, or individual, experience (in some ways echoing the experiences of the early solitary travel writers who 'discovered' Djenné). In fact, for many tourists, the presence of a large number of other tourists, especially at the Monday Market, detracts from their enjoyment of the town.

As well as the many organised tours that stop off in Djenné for a day, a high number of independent travellers come to the town. For some, such as two young students from Belgium cycling across Mali, the presence of other tourists undermined their feeling that their travels were an adventure. They described Djenné as 'touristy', a negative attribute that encompassed their feeling that there was nothing new to discover. In order to get around this and have a satisfying 'adventurous' experience, they sought out small villages in Mali and stayed there overnight, sleeping on floors and eating with families. For them, the liminal phase of their tourist 'Rite of Passage' (Cohen, 1988) had to include a physical separation from all forms of familiarity, including language. Their compass throughout their travels was a 'Petit Futé' guidebook; however, they rejected the idea of employing a guide in person. Instead, they trusted that through serendipity they would meet people who would welcome them and give them an original or different insight into Malian culture. This required a level of trust that they were aware was risky, such as entrusting their bicycles to strangers or drinking water from unknown sources.

The element of 'danger' or unscripted adventure was, however, central to their idea of travelling, a fact that became clear in their recounting of their travels in other parts of the world. Lindholm (2008: 39) describes this behaviour as being motivated by a modern desire for self-realisation and an attempt to gain a heightened sense of who you really are by testing your physical and psychological limits. He rejects the idea that most tourists are satisfied with a staged or 'fake' experience (an idea he says that has come from Jean Baudrillard's (1988) 'Simulacra and Simulations') and states that, instead, the concept of authenticity in travel has remarkable resilience.

Similarly, Wang (1999: 350) argues that the concept of authenticity has both physical (e.g. the early use of the concept in reference to the museum) and existential bases (the human condition). He states that in relation to tourism studies, it is usually the former (physical) use of the concept that is concentrated upon, to the neglect of the latter. The concept of existential authenticity is explored by Selwyn (1996), whose study of the 'myths' created by tourism builds on MacCannel's (1992) previous work:

> One of the central elements of MacCannel's argument is that the tourist goes on holiday in order cognitively to create or recreate structures which modernity is felt to have demolished. (Selwyn, 1996: 2)

This existential authenticity can be characterised by tourists such as Betty, who tend to linger where others move on, thus gaining access to behind the scenes experiences. Unlike many tourists who spend one or two nights in Djenné, Betty spent a few weeks in the town. Adopting many strategies of fieldwork, she ensured she ate with the family who ran the hotel compound and accompanied them in their daily activities. Travelling on a small budget, she also quickly managed to negotiate cheaper accommodation by moving from the tourist quarters to the family's rooms, thereby making a move from the usual 'outside' space occupied by tourists to the 'inside' intimacy of family living. She insisted that the family refer to her as 'grandmother', establishing a symbolic kinship tie and ensuring that she was looked after and included by the family.

It seems that in Betty's mind, as in the minds of many other tourists in Djenné, there are two types of culture: 'culture in the wild' (authentic) as opposed to 'culture in captivity' (monument). Her strategy was to access 'culture in the wild' through literally walking into peoples' lives and, to a certain extent, circumventing the tourist structures put in place to guide her trajectory around the country. Unlike the Belgian cyclists, she was not on a quest for an adventure in terms of physical danger and unpredictability, but instead sought a meaningful encounter with people from a different culture. This was re-enforced by the fact that after her departure, she kept in touch regularly with the family through postcards, each time signing off as 'grandmother'. Like the cyclists, Betty made reference to the 'Petit Futé' guidebook by carrying photocopied pages of the book with her, which she constantly annotated and edited. The pages were her point of entry into the culture, although by photocopying and annotating them with comments and suggestions made by the people around her, she managed to personalise them and, to a certain extent, visually distance herself from the many tourists who conspicuously carried their guidebooks with them. Her quest for authenticity was not based on a desire for historical or cultural accuracy, but instead lay in the desire for an emotional authenticity, which seemed to be at the heart of her impulse to constantly travel around the world. As noted by Selwyn commenting on MacCannell's (1973) seminal work *The Tourist*:

> If we agree with MacCannell and others that tourists seek the authentic, we need to add that such authenticity has two aspects, one of which has to do with feeling, the other with knowledge. (Selwyn, 1996: 7)

And as MacCannell, himself, remarks:

> The touristic critique of tourism is based on a desire to go beyond the other "mere" tourists to a more profound appreciation of society and culture... All tourists desire this deeper involvement with society and culture to some degree; it is a basic component of their motivation to travel. (MacCannell, 1973: 10)

However, a tension for tourists in Djenné is the disparity between the iconic reputation of the town and the reality of the poverty they encounter:

> What people "gaze upon" are ideal representations of the view in question that they internalize from postcards and guidebooks (and

TV programmes and the internet). And even when the object fails to live up to its representation it is the latter which will stay in people's minds, as what they have really "seen". (Urry, 2002: 78)

While visitors to Djenné may feel that they are distinguishing themselves through their various strategies to access 'authentic' Malian culture, Djennenkés tend to conflate categories of visitors to the town: tourists, anthropologists, archaeologists, film makers, writers, development workers, doctors, Peace Corps volunteers and so on, are all labelled as 'toubabou' (white people). One type of 'toubabou' can and does regularly turn into another, so a tourist may return as a film maker, an archaeologist may return as a tourist and a Peace Corps volunteer may return as a development worker. Despite Djenné's international cultural status, many Djennenkés believe that a secondary reason for people coming to the town is due to its poverty and so help is solicited from tourists as it is from health workers.

Large groups of tourists who come to the town are labelled by the derogatory term 'chumpas' by the guides. The tourist season is known among the guides as 'la chasse aux chumpas' (the chumpas hunt). Tourists are, therefore, commodified and seen primarily as providers of income or material help (see also Theodossopoulos, this volume). There is a genuine desire among Djennenkés to forge long-term relationships with people from other countries, mostly as a strategy out of poverty. However, many of the briefly made acquaintances in Djenné are not followed up by the promised photos or letters, leading to disappointment and resentment amongst Djennenkés.

The Mali Circuit

Almost all the tourists I spoke to in Djenné were undertaking some form of the traditional Mali tourist circuit of Bamako, Ségou, Djenné, Mopti, Dogon Country and Timbuktu. These places are widely acknowledged by the tour companies and travel guides as the places to visit in order to successfully 'do' Mali. The completion of the circuit is especially important as many tourists will only make one visit to the country in their lifetime. Other tourist destinations, such as the Northern town of Gao (home to the World Heritage Site of the Tomb of the Askias), are starting to change some tourist circuits, but only peripherally, as the North of Mali remains both physically inaccessible due to the road running out, and somewhat dangerous, due to ongoing clashes between the Tuareg population and the Malian government (for a short overview, see Oxby, 1996).

Unlike the cyclists or Betty, for whom the guides are the ugly face of tourism and emphasise the self-consciousness and staged experience of being a tourist, most tourists will visit Mali through the mediation of a guide. However, many of Betty's attitudes can be found in diluted forms in their discourse and expectations. For example, tour companies often employ a national guide who will accompany the tourists throughout their journey around the country (usually ten days to two weeks), as well as local guides on an ad hoc basis in towns such as Djenné. The additional use of local guides reassures the tourists that they are gaining access to 'authentic' or inside knowledge about a place. Local guides can also ensure tourists gain access, if only temporarily, to inside spaces such as houses or rooftops. The guides in Djenné tend to include a degree of personal narrative in their descriptions of the town and the tourists are reassured by the lively way in which they or their guide are received in the different spaces they move through around the town. The indigenous guide can also share information about a place in a conspiratorial way with the tourists and become the embodiment of the annotated comments about a destination.

The majority of the tourists visiting Mali come from a Western rationalist tradition, which is often at odds with many of the synchretic and superstitious beliefs found amongst their hosts. The tourists, there-fore, tend to adopt a different persona for the duration of their time in the country. Examining the element of role play present in many tourist encounters, Urry (2002: 98) describes how tourists taking part in organised package tours 'play' at being a child by letting all the arrangements relating to their daily needs be taken over by the tour company. Similarly, tourists in Mali have, to some extent, to entrust their daily needs to the guides, especially in the Dogon Country where the guides provide an interesting contrast with the guides in Djenné.

Van Beek's (2003) description of tourism in Dogon Country is contrasted with tourism in Northern Cameroon. Put simply, his argu-ment is that while in Dogon Country, the encounter between tourists and local people is largely positive and re-enforces Dogon feelings of cultural pride; amongst the Kapsiki of Northern Cameroon, however, the tourist encounter engenders negative feelings amongst local people about their own self-worth and re-enforces their desire to leave. He accounts for this by saying that it is the prime interest of the tourist that has consequences for how the host culture defines itself within the tourist encounter. In Dogon Country, tourists primarily come to experience Dogon village life (dances, architecture, animistic beliefs), thus re-enforcing Dogon pride. In Northern Cameroon, by contrast, tourists come to see the stunning scenery and, to a certain extent, avoid too much contact with the villages by keeping to certain defined tourist areas. This marginalisation of the local population re-enforces feelings of inferiority in contrast to the tourists' perceived wealth and prestige.

Van Beek states that each tourist encounter creates its own sub-culture. In Dogon Country, the tourists (many of whom have come from or are shortly about to go to Djenné via Mopti) enter an extraordinary world of villages perched on cliff tops, stunning views and stories of the mythical ancestors who used to inhabit the escarpment (the Tellem). Van Beek describes the attitudes of the Dogon guides towards tourists:

> The setting of the cliff villages is for them a backdrop to what is valued most, the cultural performances. The physical and cultural attraction of their country is not a source of wonder but a self-evident fact. Dogon view the relation with their visitors as more or less permanent, and on the whole are prudent not to rupture it: they should give value for money. (Van Beek, 2003: 269)

For the tourist, the experience of visiting Dogon Country is largely escapist, as you are lulled by the rhythm of walking through ever-changing scenery punctuated by stops in villages or to take photographs from vantage points (see Plate 3.2). During two visits to Dogon Country in 2004/2005, my guides described their services not only as pathfinders, but also in terms of cultural mediation, nearer to the role of 'mentor' described by McGrath (2005) in her analysis of guides working in Peru. They recounted warning tales of foolhardy tourists who had tried to explore the area alone only to run into difficulty. In particular, the guides explained that a tourist would not have the necessary cultural knowledge to know which parts of villages to avoid and would inevitably cause offence. A story I heard several times, told of a Scandinavian family who, unbeknown to them, strayed into a sacred part of a village and had to pay reparations to the village Hogon (priest) in the form of a black cow and a black chicken. The guides in Dogon Country, therefore, know themselves to be indispensable on two levels: firstly as physical guides, showing a path through the vast countryside, finding food, water and accommodation every night; secondly as cultural guides, negotiating access to villages and masked dance performances and steering tourists away from tabooed areas. The stories told by the guides of the flying Tellem ancestors blend in with the radical feeling of 'dépaysement' experienced by the tourists who spend their nights climbing up Dogon ladders to sleep under the stars and awake to the sight of stereotypical African villages punctuated by Baobab trees and surrounded by spectacular scenery.

Plate 3.2 View of Dogon Country

By contrast, in Djenné, the guides have a harder job of selling their services. Many tourists feel that they can explore the town alone, as they would a European city, armed with their guidebook and rudimentary map. In the end, unaccompanied tourists tend to accept a guide to avoid constant pestering more than through a desire for a cultural commentary. In Djenné, a guide sells his services very much on the basis of accessing inside knowledge, not in terms of regulating behaviour or physical safety. Tourists who have come to Djenné from Dogon Country will often be exhausted and want to use their time in Djenné as a relaxing interlude before starting their journey to Timbuktu, or back to Bamako via Ségou. Like Dogon guides, Djenné guides are very aware of providing value for money (usually assessed in terms of hours spent with the tourists) and will be disappointed if a tourist does not want to take them up on their offer of visiting an outlying village or Djenné-Djeno, the archaeological site.

In Djenné, more so than in Dogon Country, there is an issue with the accuracy of the guides' historical knowledge, as tourists may have a keen interest in vernacular architecture or the archaeological sites. The skills required from guides in Djenné therefore differ somewhat from guides in Dogon Country. The role-playing stance adopted by tourists visiting Djenné allows them to overlook the historical and factual discrepancies in the guides' discourses. Many of the tourists who visit Djenné are highly educated and will have researched the active interests they may have in Djenné's history, archaeology or architecture. What they are looking for from the guides is, therefore, often not the 'hard' facts that they can look up in their guidebooks or on the internet, but the 'soft' facts, or the 'art of speaking', a phrase used by one of the Djenné guides to describe his work.

The Guides

As the guides' understanding of tourists' expectations has become more sophisticated, the number of tourist locations in Djenné has increased from the traditional route (the tomb of the sacrificed virgin (Tapama Djennépo), the Holy Well of Nana Wangara, the house of the Chef du Village); to take in more unexpected sights, often brought about by tourists questions and interests. These include local schools, peoples' homes, agricultural practices, visiting the fishermen by the river and taking part in local sporting activities. Consequently, 'knowledge' about Djenné is, for many tourists, much less about facts and figures, and more about a certain intimacy, taken to extremes by Betty.

Tourists, therefore, have the power to re-enfranchise certain parts of Djenné's population who have found themselves excluded by an elitist and archival reading of their cultural heritage. This has been the case for some of the women artisans in Djenné who have organised a co-operative space to sell their work or for women working in the gardens who benefit materially from showing tourists around. However, these 'excursions into the ordinary' are only possible due to the wider framing of Djenné as a World Heritage Site and one of the top three tourist destinations in the country.

Countrywide, cultural tourism in Mali is not only about architecture and archaeology, but is being transformed by the international rise in popularity of Malian photography and music. Amadou and Miriam's internationally acclaimed album, Un Dimanche à Bamako, is one of many symbols of a new Malian modernity, picked up by international travellers and expatriates alike. Music festivals in Mali can house a broad Malian identity, at once being rooted in authenticity and tradition (e.g. Salif Keita is a modern day griot).

Conclusion

It is hard to say whether the architecture of Djenné and the consequent lifestyle of its residents denote for most tourists a 'performative primitive' (MacCannell, 1992: 26). From interviews, it seems that their understanding of Djenné is much more sophisticated than simple acceptance of the UNESCO or guidebook rhetoric. People repeatedly reported an uneasiness or anxiety about what was being asked of the town: to remain materially the same in order to retain World Heritage status. When asked explicitly about the viability of such a project, the majority of tourists interviewed felt that it was untenable and at some point in the future UNESCO would have to rethink its World Heritage project in Djenné. Many found the idea of imposing architectural restrictions on people patronising and wrong, while appreciating that it was in large part the architecture that drew them to Djenné in the first place.

All the tourists interviewed seemed keen to take part in what could be described as the 'UNESCO debate' in Djenné. Perhaps the new Djenné Museum (currently under construction) could become the focus for such a debate, allowing local people and tourists alike to have a forum in which to feedback their views. At the moment, assumptions are made on behalf of tourists (their desire for more sophisticated infrastructure, a regulated tourism industry, a predictable tourist calendar...), while the reality of long-term sustainable tourism to Mali may require a different focus.

Acknowledgements

I would like to thank Professor Michael Rowlands for his feedback on an earlier draft of the chapter and the ESRC for financially supporting my research.

Notes

1. For a full breakdown of costs including labour costs, see Djenné Patrimoine Informations, No. 21 Automne 2006. On WWW at http://www.djenne-patrimoine.asso.fr.
2. On WWW at http://whc.unesco.org/en/list.
3. Mali now has four World Heritage Sites: Djenné, Timbuktu, Dogon Country and in 2006 the Tomb of the Askias (in Gao) was added to the list. In 2007, Mali had its first UNESCO Masterpiece of the Oral and Intangible Heritage declared for the Yaaral et Dégal, the annual celebrations surrounding the transhumance of Peul cattle.

References

Baudrillard, J. (1988) Simulacra and simulations. In M. Poster (ed.) *Jean Baudrillard: Selected Writings* (pp. 166–184). Oxford: Polity Press.

Bedaux, R.M.A., Diaby, B. and Maas, P. (2003) *L'architecture de Djenné, Mali: La Pérennité d'un Patrimoine Mondial*. Leiden: Rijksmuseum voor Volkenkunde.

Cohen, E. (1988) Traditions in the qualitative sociology of tourism. *Annals of Tourism Research* 15 (1), 29–46.

Dutt, S. (1995) *The Politicization of the United Nations Specialist Agencies: A Case Study of UNESCO*. Lewiston, NY: Mellen University Press.

Eriksen, T.H. (2001) Between universalism and relativism: A critique of UNESCO's concept of culture. In J.K. Cowan, M-B. Dembour and R.A. Wilson (eds) *Culture and Rights: Anthropological Perspectives* (pp. 127–148). Cambridge: Cambridge University Press.

Fontein, J. (2000) *UNESCO, Heritage and Africa: An Anthropological Critique of World Heritage*. Edinburgh: Centre of African Studies, Edinburgh University.

Gardi, B. (1994) Djenné at the turn of the century: Postcards from the Museum für Völkerkunde Basel. *African Arts* 27 (2), 70–75, 95–96.

Hudgens, J. and Trillo, R. (1999) *West Africa: The Rough Guide*. London: Rough Guides.

Lindholm, C. (2008) *Culture and Authenticity*. Oxford: Blackwell.

Maas, P. and Mommersteeg, G. (1992) *Djenné: Chef-d'Oeuvre Architectural*. Bamako and Amsterdam: Institut des Sciences Humaines and Institut Royal des Tropiques.

MacCannell, D. (1973) *The Tourist: A New Theory of the Leisure Class*. London: Macmillan.

MacCannell, D. (1992) *Empty Meeting Grounds: The Tourist Papers*. London: Routledge.

Marchand, T. (2003) Devenir Maitre-Maçon à Djenné, rang professionnel laborieusement acquis. In R.M.A. Bedaux, B. Diaby and P. Maas (eds) *L'Architecture de Djenné, Mali: La Pérennité d'un Patrimoine Mondial* (pp. 29–43). Leiden: Rijksmusoum voor Volkenkunde.

McGrath, G. (2005) Including the outsider: The contribution of guides to integrated heritage tourism management in Cusco, Southern Peru. In D. Harrison and M. Hitchcock (eds) *The Politics of World Heritage: Negotiating Tourism and Conservation* (pp. 146–152). Clevedon: Channel View Publications.

Monteil, C. (1903) *Soudan Français. Monographie de Djenné, Cercle et Ville*. Tulle: Jean Mazeyrie.

Morton, P. (2000) *Hybrid Modernities: Architecture and Representation at the 1931 Colonial Exposition, Paris*. Cambridge, MA: MIT Press.

Olaniyan, T. (2003) What is 'cultural patrimony'? In N. Afolabi (ed.) *Marvels of the African World: African Cultural Patrimony, New World Connections and Identities* (pp. 23–35). Trenton and Eritrea: Africa World Press.

Oxby, C. (1996) Tuareg identity crisis. *Anthropology Today* 12 (5), 21.

Selwyn, T. (1996) Introduction. In T. Selwyn (ed.) *The Tourist Image: Myths and Myth Making in Tourism* (pp. 1–33). Chichester: John Wiley & Sons.

Singh, K. (1998) UNESCO and cultural rights: A collection of essays in commemoration of the 50th anniversary of the Universal Declaration of Human Rights. In H. Niec (ed.) *Cultural Rights and Wrongs* (pp. 146–160). Paris: UNESCO.

Turtinen, J. (2000) *Globalising Heritage: On UNESCO and the Transnational Construction of a World Heritage*. Stockholm: Stockholm Center for Organizational Research.

Urry, J. (2002) *The Tourist Gaze*. London: Sage.

Van Beek, W.E.A. (2003) African tourist encounters: Effects of tourism on two West African societies. *Africa* 73 (2), 251–289.

Wang, N. (1999) Rethinking authenticity in tourism experience. *Annals of Tourism Research* 26 (2), 349–370.

Chapter 4
Power, Culture and the Production of Heritage

DONALD V.L. MACLEOD

Introduction

This chapter focuses on cultural heritage and includes examples of built heritage, memorials to actual people and other forms of material culture. It examines the proactive production of heritage, which draws attention to the parties who are able to actively manipulate their environment in pursuit of creating something that represents their notion of their heritage, whether through original creativity or the interpretation of an object. And it places these examples into the context of tourism development.

Heritage is a concept that has undergone serious changes during recent years. At one time it was generally thought to embrace property passed down through the generations, to describe things that could be inherited: 'All property which is not forcibly taken by conquest but has been passed on by means of some contract or other is heritage' (McCrone et al., 1995: 1). However, since the late 20th century, the term 'heritage' has taken on a far broader remit: 'It has come to refer to a panoply of material and symbolic inheritances, some hardly older than the possessor' (McCrone et al., 1995: 1).

In a wide-ranging discussion of the term 'heritage', Timothy and Boyd conclude:

> What has clearly emerged by escalating the intellectual and economic profile of heritage is an expansion of the term to apply not only to the historic environment, both natural and built, but also to every dimension of material culture, intellectual inheritances and cultural identities. (Timothy & Boyd, 2003: 5)

According to Browne (1994), heritage can be classified into the following groups: natural (e.g. landscape, habitat, seashores), built (e.g. prehistoric remains, monuments, buildings) and cultural (e.g. literature, music, art, language, folklore). Indeed, it is an entertaining puzzle to ask what, in contemporary society, cannot possibly be or become heritage? There are already museums for computers, and many small volunteer-run heritage centres house the most mundane of domestic objects: the potential for something to become a heritage item is only limited by the human imagination.

Heritage becomes relevant to visitors as well as to the indigenous community, local neighbourhood, region and nation. Moreover, heritage may be deliberately oriented towards outsiders and visitors in order to attract them or to promote a particular image of the host community. Given the relevance of history and identity to communities worldwide, as well as the necessity to make money, the production of heritage is a sensitive, serious and valuable activity, but an activity that is permeated with ambiguity, complexity, superficiality, egocentricity and ethnocentricity in its practice.

For the purposes of tourism and the attraction of visitors, heritage has assumed an important role, and this will increase in value along with the growing need to differentiate one destination from another through the utilisation of culture as a means of distinction. In 1995, the British Tourist Authority calculated that 20% of all visits to tourist attractions in the UK were to historic properties, equivalent to 67 million visits per year (Hubbard & Lilley, 2000). In Pennsylvania, US heritage tourism contributed almost $5.5 billion to the state economy in 1997 (Timothy & Boyd, 2003: 10).

Tourism is playing an increasingly important role in the production of heritage. Various groups are aware that heritage in the form of attractions, centres, museums and monuments can become a magnet for tourists and consequently bring money into a community, region or nation. However, the production of this heritage, in terms of choice of focus, interpretation and representation will usually be in the hands of a few people who are already in positions of relative power (cf. McCrone *et al.*, 1995; Lowenthal, 1998).

In a discussion on power, Keesing (1981: 299) writes: 'Power, virtually all analysts agree, is a matter of relationships between individuals (or units such as corporations or governments) who exert control and those who are controlled by them'. Using a more abstract approach, Adams (1977: 388) defines power as 'The ability of a person or social unit to influence the conduct and decision making of another through control over energetic forms in the latter's environment'. Macleod (1998) explores the manifestation of power and its actual influences: these may be described as control over the physical (energetic), intellectual and social environments. Areas of control are divided into (1) primary areas such as contracts, payments and legislation; (2) secondary areas that are manifestations of the primary relationship and would include physical constraints over space and time: in the workplace or use of resources for example; (3) tertiary areas such as the social and intellectual activities of staff or citizens, including the restrictions on public speech and control over official historical accounts.

This chapter shows that control over heritage production can be imposed in all three areas described above, such as legislation on activity, restrictions on physical creation or display, and the prohibition of public expression regarding history and heritage. One of the primary sources of control is the nation state, which we see as producing and controlling the official heritage of the nation. In the examples given, various social units exercise some form of power and these include state politicians, local politicians, government agencies, the media, wealthy individuals and businesses. Power might also be amassed by groups of people in the form of a grass-roots movement or kinship networks who react against a more powerful, influential group.

Furthermore, this chapter explores the relationship between power, culture and the production of heritage, specifically looking at how cultural heritage is represented by groups in society and their ability to use their position to promote a particular aspect of their culture for the purpose of tourism or for their own advantage. The aspect of culture promoted may become recognised as purporting to represent the essence, symbol or archetype of the entire society; or it may simply be the only part of a culture that is promoted. In each case study, there are examples of particular people who have been memorialised, some-times by a statue, or through portraits, plaques, images and dedicated spaces. This public recollection of actual individuals is one of the more blatantly sensitive aspects of cultural heritage and enlightens contemporary contests over identity and ownership.

There are some overarching patterns that present themselves in the case studies and these become manifest as official, state-supported heritage in contrast to unofficial, grass-roots heritage. In general, the official heritage is utilised by the government tourism authorities for marketing purposes; however, we can see that unofficial heritage can eventually be embraced by tourism agencies as a means to attract visitors. Similarly, state organisations and representatives, such as political parties, may come to embrace the unofficial grass-roots heritage if it suits their purposes. The official, national heritage, as opposed to the unofficial, grass-roots heritage is a binary opposition that will serve as a flexible template for our analysis of the case studies.

By looking at three examples based on field research, this chapter offers a comparative study in differing socio-cultural and political environments and seeks to draw similarities, enlightening us about the place of power in the process of heritage production and representation, and its relationship with tourism.

Valle Gran Rey, La Gomera

Introduction

La Gomera is one of the seven Canary Islands located close to the west coast of Africa. They form an autonomous region of Spain. The islands were inhabited by indigenous people known as the Guanches, believed to have settled some 3000 years ago, arriving from Africa and of Berber origin (Hernandez-Hernandez, 1986; Castellano-Gil & Marcias-Martin, 2002). The Guanches pursued a pastoral economy using goats, sheep, pigs and cattle; they had a stratified society with a system of kingship as well as an elite tier of religious leaders. The Iberian powers conquered the islands over a long period during the 15th century, coming to different arrangements on the separate islands, which had hitherto operated independently.

The Guanches form a central part of the modern Canary Islander identity, manifest through activities such as eating and drinking, work and leisure pursuits, and recorded in literature, art, historical records and popular writing such as Concepcion (1989). It is believed that numerous traditions and skills have been passed down from the Guanches, which form part of contemporary culture including musical instruments, dance, song, household items, pottery, basket-weaving, herbal medicines, animal husbandry, the Silbo whistling language, food and drink (cf. Galvan-Tudela, 1995). These phenomena may be part of everyday modern life today or celebrated in museums, exhibitions and festivals.

There are many day-trippers from Tenerife visiting La Gomera on coach tours arriving via sea-ferry. The coaches disembark at San Sebastian, the capital and port, and drive around the island stopping at the visitors' centre in the middle of the island (where Guanche traditional skills are used to make items for sale) as well as the coastal destinations: Valle Gran Rey and Santiago. Tourists are predominantly German, including independent backpackers and (more frequently since the arrival of the internet) those who have booked self-catering apartments in advance: they dominate the winter season. In summer, the Spanish holidaymakers arrive, mostly from Tenerife and mainland Spain. By 1999, there were 600,000 visitors to the island in total, and 5000 beds available on the island, with 3500 in Valle Gran Rey alone.

Tourism has come to be the major industry for most of the larger islands (Tenerife, La Palma, Gran Canaria, Lanzarote) and forms a large part of the Gross Domestic Product (GDP) for the region. There were 10 million foreign tourists visiting the archipelago in 2001 and official data describes the tourist sector as playing a central role in the economy (Pascual, 2004). Tourism has become increasingly important for La Gomera, where the primary industries of agriculture and fishing dominated until the 1980s; this small island has a population of around 22,000 (2006) and an area of 378 km².

The commodification of culture: Spanish national heritage

La Gomera is also known as *La Isla Columbina* – Columbus Island. However, this appellation is rarely used by the islanders themselves, but appears in tourism promotional material: brochures refer to his residence on the island. It was here that Christopher Columbus repaired his ship and prepared his crew for their initial journey across what we now know as the Atlantic Ocean: San Sebastian, La Gomera, was his final departure point. He is associated with numerous buildings in the port and capital San Sebastian, where he was based for several months. One house where he stayed is now a museum; another building is the *Torre del Conde*, former tower house of the Count of Gomera (cf. Bianchi, 2004).

Today, Columbus is an international icon, a very strong brand for marketing purposes, and strengthens the island's promotion of its historic built heritage and role in the Spanish colonial experience. A recently sculpted bust of Columbus now stands outside the Tourism Information Office in San Sebastian and there is a large gallery in the building, which celebrates him with biographical details, models of his ship, maps of his journeys and depictions of contemporary island life.

Along with the association with Columbus, elements of the culture of the pre-Hispanic indigenous people (The Guanches) have remained among the population, and skills such as the whistling language (*Silbo*), basket-weaving, hand-made pottery, music-making and dance have become commoditised and are sold to visitors through material gifts or attractions (see Macleod (2006) on commoditisation). Examples of such gifts include hand-held drums (*tambors*), large castanets (*chacarras*), hand-made pottery, and figures and symbolic designs based on pre-Hispanic

Guanche belief. The visitors' centre (known as *Juego de Bolas*) in the middle of the island hosts a model of a traditional peasant's home, and has various people employed on the site to demonstrate how traditional craft products are made. It also hosts a model of Columbus's ship (Santa Maria), and an exhibition of regional flora and fauna. Some of the Spanish Roman Catholic religious festivals, such as that celebrating the Virgin Carmen, have been commercialised and have become extended periods of late night entertainment and profane events (such as donkey races) lasting up to five days. They also offer cultural entertainment including traditional (pre-Hispanic) *tarajaste* dancing and Gomeran music.

In contrast to the above examples of the commodification of culture and its transfer into the market-place by official organisations (such as the *Cabildo* – Island Government – and the Gomera Island Tourist Board), some of the local people in the popular tourist destination of Valle Gran Rey (a municipality on the southwest coast) lament the passing of simple festivals that focused on the original religious meaning, and miss the collective community actions associated with celebration such as the burning of bonfires on the festival of St Mark. They worry about the loss of their access to the beach because of overcrowding by visitors, the lack of time to spend with their families because of changing work patterns and increased hours, and the diminution of fishing as a livelihood.

The decline of fishing and associated heritage

Fishing has been an important economic activity for the people of Valle Gran Rey for more than 100 years. In earlier times it formed part of a number of ways of surviving, including subsistence farming and plantation work, and was a good means of obtaining additional protein. During the 20th century, the port of Vueltas, on the coastal margin of the valley, developed into a busy fishing village with a population of over 300 people almost wholly dependent on fishing as a livelihood. However, since the 1990s, there has been a rapid decline in the number of professional fishers based in the valley. This economic livelihood had a major influence on the culture of those associated with it, impacting on kinship alliances, marriage partners, social activities and perspectives on the world: the decline of fishing has meant the erosion of numerous social and cultural phenomena associated with the fishing community (see Macleod, 2002).

People in the fishing community believe their heritage is being eroded, physically disappearing or beyond their grasp, and not commemorated.

This is a less obvious sort of tragedy than the dramatic transformation of a cultural event exemplified by the Alarde of Fuentarrabia, as documented by Greenwood (1989) and others, where a historically important festival is turned into a show; rather, the transformation of Vueltas is a more profound and widely felt development. These transformations witnessed on La Gomera, moving from a fishing and agricultural lifestyle to one oriented towards the tourism sector, reflect the experiences of many people worldwide living in tourist destinations, where livelihoods once based on primary forms of production are disappearing, leading to a reorganisation of their social and cultural lives (cf. Boissevain & Selwyn, 2004).

The transformation can be very rapid: for example, Vueltas has changed in one generation from being a traditional fishing village with a small harbour full of fishing boats into a tourism and recreation centre with a port and harbour dominated by leisure boats and ferries. Valle Gran Rey was composed mainly of private dwellings, smallholdings and banana plantations up until the 1980s; the agricultural sector has been reduced since then, while many cultivated terraces have been abandoned. Old vernacular houses have been demolished, especially in the fishing zone, the original village rising vertically as additional storeys turned into tourist accommodation.

Monuments and local history: The people's heritage

Two significant monuments to people and events involving the local population have been established on the coastal plain of Valle Gran Rey since the mid-1990s. The first one, on the edge of Vueltas, is located on a large, grassed roundabout (funded by the European Union) built in the mid-1990s. The memorial commemorates the sailing boat 'Telemaco', which illegally transported 171 people from the valley across the Atlantic Ocean away from poverty in a quest for economic survival in Venezuela. The memorial is composed of an 11m fishing boat named Telemaco, once owned and used by a fishing family in Vueltas for regular work and offered as a donation to preserve the memory of the original journey. There is also a large boulder on which a metal plaque describes the event that occurred in the 1950s during the period known by Canary Islanders as *La Miseria* (the misery). This was a time of poverty and desperation, when Spain suffered hardship and lacked support from the victorious allies after the Second World War; consequently, people were obliged to remain in Spain to develop the economy, and emigration from the Canary Islands was severely restricted. A film entitled 'Guarapo' depicts this famous episode when people risked their lives crossing the ocean, often in inadequate vessels, including the Telemaco.

The second monument is a statue measuring around 4m in height, commemorating Hautacuperche, who is depicted carrying a broken bowl in one hand and a spear in the other (see Plate 4.1). He was one of the leaders of the 'Rebellion of the Gomeros' against the Spanish colonial authority on La Gomera in 1488. The Count of La Gomera, known as Hernan Peraza the Younger, was the Spanish representative and colonial master on the island. The rebellion began because the Count was regarded as a tyrant who had broken an agreement between the Spanish and the Gomeros. Moreover, he had conducted an illicit relationship with a Gomeran woman known as Iballa. Hautacuperche is celebrated as the man who entered a cave and killed Peraza, but who was himself killed in a second battle against Spaniards protected within the *Torre del Conde*, the tower house of the Count in San Sebastian. The murder of the Count led to heavy-handed repercussions from his surviving wife, the Countess Beatriz de Borbadilla, involving violence towards, and the enslavement of, the Gomeros.

The slaying of the Count of La Gomera, a historical and politically momentous event is celebrated throughout La Gomera, and is particularly close to the people of Valle Gran Rey, where the conspirators met to plot against the count and where they would pass information while meeting

Plate 4.1 Hautacuperche

on a small rock located some 200 m off the coastline of Vueltas, known today as the 'Rock of Secrets'. These meetings and the entire episode have been recounted in stories and song passed down through the generations.

The imposing statue of Hautacuperche was erected on Canary Islands Day (30 May) 2007: a plaque commemorates the event with the Canary Island Government, the Island Council of La Gomera and the Regional Council of Valle Gran Rey giving their support. The establishment of this statue has been promoted by a political grouping 'The Canary Islands Coalition', composed of political parties including the Central Democratic Party, which happens to be the majority party holding power in the municipal council of Valle Gran Rey since the late 1980s. It was suggested by one local resident that these politicians are seeking to show an allegiance to the people, as well as promoting the raising of consciousness regarding the distinctive identity of the Canary Islands in relation to mainland Spain.

A Canary Island Government sponsored leaflet explains:

> This monument is a testimony to the bravery of the Gomero people, it is in recognition of their nobility and courage and a tribute to the force necessary to arrive at what is today La Gomera: a free community, wealthy and prosperous, with solidarity and the capacity to learn from the past in order to live better in the future. (My translation)

It is interesting to note that both monuments discussed above celebrate the local people and their attempt to avoid oppressive circumstances precipitated by the state (cf. Bianchi, 2004). Both events are well known by people living in Valle Gran Rey and form part of the folk memory as well as formal, official history. People in the valley identify directly with the events whether through considering their own Guanche past or remembering the impact of *La Miseria* on their lives or those of their relatives (see Macleod, 2004a). This local history linked to monuments contrasts directly with their association with Columbus, which is minimal. Columbus could easily be regarded as a representative of the state, and he is said to have had an amorous relationship with the Countess Beatriz de Borbadilla, an orchestrator of oppression for the Gomeros.

There is, therefore, a massive contrast between the cultural heritage celebrated by monuments in San Sebastian and that celebrated in Valle Gran Rey. San Sebastian boasts of its association with and accommodation of Columbus, who departed the port for his first voyage across the Atlantic. Along with the colonial buildings, a large mosaic in the recently redesigned plaza depicts his route across the Atlantic. Marketing literature has promoted the connection with Columbus and every year, festivals celebrate the international connections that have grown since his journeys.

Whereas, Valle Gran Rey monuments recollect the brave journey of Gomeros across the Atlantic Ocean, away from poverty and oppression during the Franco era. They also commemorate the popular rebellion against the early Spanish colonisers. We might explain this contrast by considering that San Sebastian, as capital and the main seaport of La Gomera, is celebrating its Spanish colonial and maritime heritage and international links, especially with Latin America. Columbus played an immensely important part in the development of the Canary Islands as a trans-Atlantic link between the old and new worlds (Fernandez-Armesto, 1991). He is an icon, a globally recognised figure, and as such a useful device for gaining prestige and recognition. This means he can be used advantageously for publicity, not least in tourism terms. The use of Columbus is part of a marketing ploy, which looks outwards towards the international community, potential visitors and friends of the islands.

In comparison to San Sebastian and the Columbus connection, Valle Gran Rey is celebrating local heroes. The monuments commemorate the history of the people and are inward looking in that they seek to address indigenous Gomeros, and may serve to bond these people together and to their collective past. The two monuments in Valle Gran Rey do not immediately seek to attract the attention of outsiders, potential visitors. However, they do have the capacity to emphasise the unique history and experience of the local people, something that is of increasing importance in the globally competitive world of tourism where destinations need to establish distinct identities to help attract tourists. It may be a coincidence, but the statue of Hautacuperche now appears as an image on leaflets promoting island culture in the valley tourist office.

A division between the heritage of Spain as nation and that of the Canary Islands, an autonomous region, is visible when we consider the examples above, where the rebellious Hautacuperche and the desperate refugee migrants in the Telemaco sought to challenge the might of the state through direct opposition or escape. The Spanish nation state would not wish to celebrate those who have fought against it, or tried to escape illegally: as a primary area of control, the government is able to restrict the public communication of such events. However, given the historical relevance, the distance in time and other factors, it clearly does not see these monuments as sufficiently threatening to be discouraged. Certain political representatives have chosen to support the memorialisation of these events and symbolically embrace them as distinct representations of the rich history of the place. Similarly, tourism marketing material is beginning to use them as examples of the cultural distinction possessed by La Gomera.

Bayahibe, Dominican Republic

Introduction

The following case study of the Dominican Republic deals largely with heritage that relates to the representation of national identity that has been defined (to an extent) in opposition to its neighbour Haiti with which it shares the same island. An example is given of how heritage can be utilised to create a sense of group unity and establish boundaries for the group. A strong sense of history, which has been manipulated to suit powerful elites, is also apparent. In contrast to the national heritage described, the study also examines a grass-roots drive to achieve a sense of local identity, belonging and heritage, through ownership of the land using legal contest, memorialisation and the recording of historic events.

The Dominican Republic shares the same island with Haiti and occupies almost two thirds of the territory, with a population of approximately nine million. Hispaniola was the name given to the entire island by the Spanish after colonisation following its discovery by Columbus in 1492. It had previously been discovered by Amerindian groups thousands of years earlier and they were occupying it when the Spanish arrived: one ethnic group, the Taino, were dominant at that time. Santo Domingo, the capital, was founded in 1496 by Bartolomé, the brother of Christopher Columbus. By the early 16th century, the Amerindian groups were being severely repressed and almost eradicated through illnesses and enslavement. Eventually, slaves from Africa were imported to work on the plantations.

In 1697, the western part of the island was ceded to France and became known as Haiti. By 1795, France had gained nominal control over the entire island, but by 1804, Haiti declared independence, the first example of its kind. In 1822, Haiti governed the whole island, although it was eventually defeated, and by 1844, the Dominican Republic declared independence; Spain briefly regained control, but independence was recovered in 1865. Thereafter, until the 1960s, much of the governance of the Dominican Republic had been through dictatorships, with the US military occupation occurring in 1916–1924 and military intervention in

1963. The Dominican Republic is currently a representational democracy (Howard, 1999).

This brief review of historic political events illustrates the fraught relationship the country has had with its neighbour Haiti, one that is further exacerbated by ideological differences relating to cultural and ethnic heritage, as well as serious economic disparities between the two nation states, with Haiti experiencing extreme poverty and the Dominican Republic gradually becoming wealthier, partly because of income from tourism.

Currently, the Dominican Republic is one of the most popular tourist destinations in the Caribbean, receiving US$1.8 billion in 1997 (Howard, 1999). Other sources of income include agricultural and mineral exports, manufacturing and remittances from overseas ex-pats. Despite the dramatic growth of tourism income in recent years, the country remains relatively poor with a GDP of US$9200 per person in 2007.

Official cultural heritage

In his first journey of exploration, Columbus left La Gomera and travelled across the Atlantic, eventually reaching the island of Hispaniola. In time, the Spanish established a colony on the island, and the capital city, Santo Domingo, contains a large area of Spanish colonial architecture preserved and presented as the *Zona Colonial*, which has been designated a World Heritage Site. The zone contains a park dedicated to the memory of Columbus, with a large statue of the famous explorer as well as the first cathedral built in the Americas, a church, a fort, houses belonging to wealthy individuals, royal buildings, a house for Jesuits, a palace complex, the residence of the son of Columbus who became the first viceroy of the New World, a Franciscan monastery, a hospital, town hall, convent for Dominican followers and numerous other historically important buildings and parks. These all testify to the strong links that the Dominican Republic had with Spain, a connection continued in the official language (Castilian Spanish) and religion (Roman Catholicism). They also form a large part of the 'product' promoted in tourism literature: for example, the *Insight Compact Guide to the Dominican Republic* (Latzel & Reiter, 1998) devotes several pages to the subject of the Colonial Zone.

An elite cadre of businessmen and politicians who occupy the upper echelons of government and the establishment have promoted the Spanish heritage of the country and its people, sometimes unfairly overshadowing other groups and ethnicities who have contributed, specifically those of African descent and the original inhabitants, the Taino Indians (Pons, 1997: 243; also see Calvo-Gonzáles and Duccini, this volume, for a similar phenomenon in Brazil). One example of the state officialdom making a deliberate statement about the nation's cultural and ethnic roots is seen in the developing form of the *Museo del Hombre Dominicano* (The Museum of Dominican Man) in the capital city, Santo Domingo. Here, the state had, at one time, represented its cultural heritage through two huge statues standing outside the entrance: these were of the Spanish priest Bartolome de Las Casas and the Amerindian 'Taino' chief Enriquilo. However, there was one very important group missing: those of African descent, many of whose ancestors were transported to the Dominican Republic from Africa across the Atlantic as slaves. This deliberate exclusion was eventually rectified and a statue to Lemba, the ex-slave leader and symbol of emancipation was erected. As an example of the manipulation of heritage by a powerful group, this is exemplary. The governing elite have maintained and promoted their cultural links with the early Spanish colonisers and Hispanic culture. They remain a dominant economic, political and cultural force in the country, and have previously sought to organise their history and heritage very tightly to reflect their prejudices (see Dobal, 1997; also Lowenthal, 1998, on the bias of historical works and heritage).

The scholar, Frank Moya Pons, makes the point that in the early days of the Dominican Republic, it was a minority who controlled education and communication because they were able to read and write and had the unique capacity to leave documents for posterity; this minority also governed the country. Furthermore, he notes that: 'In all Latin American societies the conquistadors imposed an order of things that obliged the population of indigenous Indians, negroes and mixed races to accept the whites as excellent' (Pons, 1997: 238). Dominicans referred to themselves as 'whites of the soil'. This goes some way to explain the governing elite's antipathy towards the Haitians who were proud of their African slave origins. This mentality has continued among many members of the establishment and an affinity towards Latin American countries as opposed to the Caribbean grouping, as well as hostility towards Haiti, was noticeable in some newspapers during electioneering in the Dominican Republic in the year 2000 (Macleod, 2005).

The production of heritage and representation of a population or place through specifically chosen icons and symbols also occurs in the maritime village of Bayahibe on the southwest coast, with a view to attracting visitors. This small village, recognised by many locals as being established in 1798, is represented and marketed by national organisations as a 'fishing village' in brochures and on billboards using images of the picturesque gaff-rigged sailing fishing boats that have been associated with the area. This image is becoming increasingly misleading as the fishermen have dwindled from a total of around 100 in the 1980s, to some 15 in 2004. Most have found work on the motor boats that ferry tourists to a small island: Saona, part of the National Park Del Este. The fish have been driven away from the coastline by the heavy traffic of motor boats, necessitating the few remaining fishermen to travel much further for their catch, increasing their fuel costs and time spent at sea. In short, the attractive image of the 'fishing village' is rapidly becoming wholly inappropriate to reality in Bayahibe as it develops into a transit port for package tourists on their way to the pristine beaches on the small island of Saona. Very few tourists enter the village to purchase goods or stay in accommodation: rather, they are shepherded onto waiting motor boats and driven off immediately.

Grass-roots cultural heritage

From the official state representation of the national cultural heritage and the official image-making and branding of a 'fishing village', we move to unofficial local 'grass-roots' understandings, interpretations and cultural representations of community and family heritage. One family, the Britos, is busy writing up its history and has made claims to ownership of Bayahibe village land and surrounding plots in the courts (Macleod, 2005). They believe that their ancestor, Juan P. Brito, who arrived from Puerto Rico, had several children, and the youngest, a son, bought the land, but he had his title to the land stolen many years later by the local government. They have nailed the ancestor's portrait to a post in the village square and celebrate him as the 'Founder of the Village'. This portrait of the 'founder' is joined by another oil painting nailed to the post: that of one-time politician, Senator Alberto Giraldi, a native of France and a politician with the Dominican Revolutionary Party (PRD), elected as Senator for La Romana province, which includes Bayahibe. He defended the villagers in their struggle for the rights to the land, which they believed to have been wrongly taken from them. He is described as 'The father of the liberation of Bayahibe' in a hand-painted description next to his portrait (see Plate 4.2). This 'unofficial' creation of a special place is a physical manifestation of cultural heritage, a form of 'representational space' as distinguished by Lefebvre (1991) in the sense of space created by local inhabitants through their daily lives

Plate 4.2 Heroes of Bayahibe

and experience: as opposed to one created by government planners (see Macleod, 2004b: 40–41).

Together with the written history of the Brito family in the region, the commemorative representational space begins to engender a strong sense of heritage among family members, which is disseminated to other villagers. It is a narrative of the past and an explanation of the present situation in the village that inhabitants hear regularly and interact with. For the Brito family, the land is their heritage, as is their family history, both of which are indigenous productions, and they are not sanctioned by the officials of the nation state. In theory, it might be possible for the Brito family to create a museum commemorating their family history and that of the village, and we could imagine this becoming a tourist attraction; however, a lack of resources at least means this is unlikely.

The culture of the villagers, including family life, local history, music, dancing, religion, cooking, farming and fishing is overlooked as a resource for attracting visitors. This contrasts to stereotypical aspects of 'Caribbean' culture that are promoted in brochures and sold to tourists as entertainment in nearby hotels where evening performances are given, which include rum, calypso, reggae and merengue. The closest that many hotel guests get to the villagers is through helicopter trips over the village and quad-bike rides through it. The hotels actively discourage their guests from leaving the hotel grounds unaccompanied (this corresponds to findings recorded by Sommer and Carrier, this volume).

In this example, it seems that the world of the villagers, their own sense of family and village heritage is very distant from the cultural heritage promoted by the state: the division between official and unofficial heritage is clear. There is a sense of a muted celebration of freedom from oppression by the state in the case of the Brito family. Nevertheless, there is a common theme promoted by the state, which can see commercial advantage to utilising the fishing heritage and tradition of the village. And yet, the reality is that the villagers and other business interests are busily, albeit indirectly and through necessity, marginalising the fishing economy in this small community. The power of the tourism industry has led to the diminution of the fishing in the village, and at the same time encouraged the government authorities to brand Bayahibe as a fishing village: an ironical and unsustainable outcome. Meanwhile, the relatively poor villagers struggle to retain their historical links to the land through aggressively pursuing their legal rights, recording their history and representing their ownership through visible, if fragile, public memorialisation. Recognition of the local heritage as rich and distinctive has not yet been made by the tourism industry.

Dumfries and Galloway, Scotland

Introduction

Dumfries and Galloway occupies the southwest corner of Scotland, and extends some 90 miles east-west and 45 miles north-south, containing a population of around 140,000. It has a rich history, embracing visitors, invaders and occupiers from a variety of ethnicities including the Celts, Angles, Britons, Irish-Scots, Romans, Normans, Vikings and the English; all have left physical remnants on the landscape and cultural elements among the people (see Robertson, 1992). There are stone circles, Roman roads, medieval castles, ruined abbeys and examples of fine architecture to be found throughout the region. Writers, poets, artists, engineers, philosophers, scientists, bankers, inventors have all been illustrious inhabitants at some time; one local man became the founder of the US Navy. Its proximity to England invited many major battles, sieges and the building of castles and fortified homes; it also led to the development of a no-man's land separating the two countries, a zone known as the 'debateable lands', where the infamous 'Reivers' undertook to steal cattle, amongst other nefarious activities.

Nevertheless, despite its rich cultural heritage, the region is currently considered to have problems establishing a strong identity in modern times by agencies such as VisitScotland and the local council. The local tourist board helped develop the strap-line 'The natural place to be', which whilst being pleasant, is undoubtedly insipid and indistinct, and might easily be applied to other destinations (Nepal uses 'Naturally Nepal'). Tourism brochures have certainly described the more memorable, dramatic historical events in the region, but a genuine capitalisation on the region's cultural heritage does not yet seem to have occurred. Powerful groups of people, and groups struggling for power have left their marks across the landscape through built heritage and archaeological remains, but many visitors to the region are unaware of the rich history in their midst, and arrive for the purpose of relaxation, and more recently, vigorous outdoor activities such as mountain-biking: the attraction of natural heritage is strong.

There is a large gap between the assets that the region possesses, its comparative advantage, and the way it uses its resources, its competitive advantage. This section shows how unofficial groups have begun to use the cultural resources of the region to promote particular places, especially towns. Whereas the well-known Scottish icon, Robert Burns, continues to be vigorously promoted by national organisations, unfortunately the region lacks clear and total association with him and becomes hindered in its objectives.

Tourism is very important to the region in terms of employment and income; it receives around one million tourists per year contributing almost £150 million, and in addition 11 million day visitors (VisitScotland, 2006; TNS, 2006). It is estimated that 11% of the working population are involved in tourism-related services, slightly higher than the national average for Scotland. It is, nevertheless, regarded as an agricultural region, providing a substantial proportion of Scotland's cattle and sheep; it is heavily forested and has a light manufacturing industry based in Dumfries.

Robert Burns: Scotland's National Poet

Robert Burns spent his final years in and around Dumfries before his premature death in 1796: he had chosen to live in this region, as opposed to his region of birth, Ayrshire. He had worked as a customs and excise officer and was an active member of the local defence corps and Masonic club, as well as being a man of letters and song. Burns was famous in his own lifetime and had a wife and lovers who bore him many children. He travelled widely in the country and there are sites that claim association with him throughout Scotland. In Dumfries alone there is a Robert Burns Centre; his old town house – now a museum; the Globe Inn, which keeps as an exhibit, a bedroom where he stayed, as well as a dining room devoted to his memory, and his mausoleum in St Michael's Church graveyard. Just outside the town is Ellisland Farm, where Burns tried to make a living as a farmer: it now retains his living quarters as an exhibit. Dumfries possesses a 'Burns walk' along its river (The Nith) and has an impressive statue of him with his dog at one end of the town, while at the other end, opposite St Michael's church, stands a recently erected statue of his wife, Jean Armor. One taxi firm sports his facial image on their cars; and one man offers a private 'Burns Tour' of associated places in Dumfries. There are, in short, a plethora of physical associations with Burns throughout Dumfries and its proximate region, and the tourism authorities regularly mention him in their marketing media: for example, the brochure 'Dumfries and Galloway See and Do 2006' produced by VisitScotland Dumfries and Galloway, devotes an entire page to him as 'Scotland's National Bard', describing his life, and listing the opening times of his home, the Globe and the Burns Centre (D&G, 2006: 22).

However, there is one problem regarding Burns in Dumfries, and that is Ayrshire, where Burns was born and where the Burns National Heritage Park is based. This unspoken competition means that Dumfries is certainly not the monopoliser of Burns and his memory, and has, to some extent, been the weaker candidate as a visitor attraction. Nevertheless, because of the strength of support for Burns due to his national and international fan-base, he remains an icon and deemed worthy of commemoration. Only recently, a national competition known as 'Burnsong' was held in schools celebrating his legacy, and furthermore in 2009, a 'Homecoming' event will celebrate 250 years since his birth, on an international scale.

Robert Burns is a national icon and has been claimed by numerous places around Scotland as someone with whom they have a strong association. There are many sites in Dumfries that have strong links with Burns, yet despite the critical mass of highly important sites, the town and region do not attract large numbers of visitors solely through the poet and related cultural heritage. This is partly because of the strong competition from Ayrshire. Powerful organisations funded by the Scottish Government continue to market Burns in Dumfries, and he is an example of the state-sanctioned national identity used as a heritage attraction.

The theme towns: Grass-roots developments

In contrast to the focus on Burns, there are local grass-roots initiatives celebrating cultural heritage, which are not directly promoted by public-funded organisations. The theme town developments are examples of such initiatives: an interesting and recent phenomenon in the region.

Plate 4.3 Castle Douglas Food Town Festival day

These are initiatives largely undertaken by the indigenous population, sometimes with the stimulus from national agencies that have created new structures of cultural heritage, for example: Wigtown Book Town, Kirkcudbright Artist's Town and Castle Douglas Food Town (see Plate 4.3). These towns are building on their association with products, activities and people, and have evolved their proactive branding since 1998 when Wigtown was awarded the title of the first Scottish Book Town. Much of the work has been initiated by volunteers; thereafter opportunities arose for funded posts to employ town development officers. Specific bodies have steered and managed the towns' developments, which have included book festivals, the refurbishment and organisation of an old town hall building, art exhibitions, a campaign for a national art gallery outpost and annual festivals.

Wigtown itself has experienced a growth in the number of businesses in the town, a substantial increase in property value and an increase in the number of visitors, all directly related to its book town status. Kirkcudbright has experienced a growth in its visitor numbers linked to its regular exhibitions. This success is stimulating other towns to build on their heritage and promote themselves based on a local characteristic: thus Moffat, once a spa town, is promoting itself as a centre of 'Wellbeing', and Newton Stewart is emphasising its links with outdoor pursuits. Another town, Dalbeattie, has considered using its associations with granite quarrying and building. All hope to increase the numbers of visitors into their towns and encourage spending in the vicinity.

Authority and the presentation of cultural heritage

The above towns are promoting products that generally fit into the definition of 'culture' as understood by tourism agencies: a form of 'high' culture supposedly appealing to the educated audience. The concept of 'culture' has been interpreted by VisitScotland (ex-Scottish Tourist Board) in tourism brochures and development strategies as the following: visual arts, literature, film and music (DGTB, 2001); 'cultural products' as listed include festivals and events and castles (DGATP, 2007). This approach tends to ignore other aspects of culture as understood in its broadest sense by anthropologists (e.g. Geertz, 1973; Tylor, 1871), which include different modes of livelihood, folk beliefs and customs, everyday activities and material items: the type of material culture that forms the bedrock of many local heritage collections around the world, and is to be seen increasingly in Dumfries and Galloway (e.g. the small towns of Creetown, Dalbeattie and New Galloway), manifest in collections of photos, household products and community memorabilia (c.f. Lowenthal, 1998: 3).

One recent example of unofficial, grass-roots development is the 'Wicker Man' music festival, which is very loosely based on a 1970s film made partly in the region. This festival began as an outdoor music festival in 2002, run by a farmer on his own land. It received a lot of press attention in the region because *The Wicker Man* film itself was regarded as anti-Christian by some people, and the festival was consequently roundly criticised for giving the region a bad image. In its early years, it attracted a few thousand people, a figure that grew to just over 10,000 by 2007. The festival has received some financial support from the council, and due to its continuing success is now being promoted in tourism literature. It represents the division, albeit hazy, between officially promoted cultural heritage and grass-roots, unconventional unofficial heritage. It also shows how non-mainstream heritage can become adopted by officialdom relatively quickly.

Nevertheless, we can conclude that there continues to be a representation of the cultural heritage of Dumfries and Galloway by 'authority' (state power) such as official agencies including VisitScotland, in terms of icons and 'high' culture: with such notables as Robert Burns the poet, together with images of castles (see the brochures: D&G, 2002, 2004, 2006). Such ubiquitous icons may equally be used to represent the nation as a whole and inadvertently veil the region's local distinctiveness. A similar approach can be seen with the National Trust for Scotland: McCrone *et al.* (1995) show how traditional and conservative the membership of this organisation is, as is its definition and approach towards heritage. It is an organisation that is dominated by people from the more powerful sectors of Scottish society: the authors note 'From the outset, this organisation has had a strong aristocratic and landlord domination of its council' (McCrone *et al.*, 1995: 101). In a time when tourism destinations need to distinguish themselves and their unique selling points, the narrow focus of some Scottish agencies can become an expensive oversight.

The division between the official agencies of the nation and unofficial grass-roots initiated promotion of heritage is not absolutely clear in this case study. Many of those promoting theme towns are contributing their time voluntarily while holding jobs in local government or quangoes. Some towns have received seed-funding from the local or national government. However, in the majority of cases, it is the local people living in the towns, who have done the actual work and initiated and supported the projects, as opposed to centrally controlled agencies like VisitScotland or Scottish Enterprise taking an official lead. Moreover, it must be concluded that it is those local people with cultural capital (Bourdieu, 1986) in the sense of formal education, professional experience and network contacts, and having discretionary time, who are able to take a lead in the development of theme towns.

Conclusion

Power is exercised variously by different groups in accordance with their ability to enact their desires. There is a clear link between power and the promotion of heritage, especially for the purposes of tourism, as only those able to do so can express, manifest and legitimately promote their heritage on a large scale, significant enough to attract the attention of potential visitors, or to influence the wider public. For example, in the Dominican Republic, officials of the government influenced the interpretation of the nation state's ethnic cultural heritage as represented in the official Museum of Dominican Man and originally ignored the African ethnic group in its symbolic statues outside the entrance; while the original Spanish colonial buildings remain protected to an extent and promoted through their World Heritage Site status. This contrasts with the Brito family's relatively miniscule and fragile commemorative portraits of the 'village founders': a cultural heritage space constructed to inform villagers and possibly visitors, with one of their intentions being to reinforce their claims to the territory.

Similarly, within La Gomera, government agencies help maintain the dominance of the connection with Columbus as well as the museumification of the Guanche culture and peasant activities that represent the island's cultural heritage and offer specific marketable products. By contrast, aspects of recent cultural heritage, for example fishing traditions, are being overlooked, while religious ceremonies and festivals are becoming increasingly commercialised. Meanwhile, people and events that represent acts against locally perceived state repression have, only recently, become officially memorialised, such as the Telemaco and Hautercuperche; but they are not promoted to visitors as attractions.

Within Dumfries and Galloway, government agencies such as Visit Scotland continue to promote Scottish icons including Burns, and built heritage, including castles, which dominate the promotional media: they are directing and constructing a particular 'tourist gaze' (see Urry, 1990). However, grass-roots desires to represent folk heritage and very recent memories is growing and beginning to create a new cultural image for the region. This is particularly apparent with the creeping success of the theme towns.

These examples have shown that there has been a gradual assimilation of grass-roots cultural heritage into various nation states' representation of their cultural experience and heritage. The African heritage in the Dominican Republic; the rebellious Gomero Hautacuperche, and illegal Gomeran emigrants on the Telemaco; the theme towns and the Wicker Man Festival in Dumfries and Galloway: all have become part of the representations promoted by state agencies. Where tourism enters as a consideration into the equation, the state, through its agents, becomes especially active in promoting identities and facets of history that are deemed attractive. Thus, we perceive the irony of the promotion of Bayahibe as a fishing village at the same time as its fishing industry disappears. Heritage promoted for the purpose of being consumed by tourists might be seen as superficial and possibly irrelevant to those indigenous people experiencing contemporary events. Columbus is largely irrelevant to the current inhabitants of La Gomera: identities are promoted beyond the locale, which may be out of joint with the destination's inhabitants and their worldview. Economic forces are driving the image making, operated by powerful business and political interests.

Statues and other portrayals of actual people become important and fascinating parts of this cultural heritage process because they are representative of those who once lived, and they become emotional touchstones (sometimes literally) for the local populace and others.

A statue becomes a symbol of a person or a group of people. Because of their intrinsic power to represent and epitomise someone or something, statues are also vulnerable to symbolic destruction, especially those with religious or political significance. When political regimes replace others, they will often remove the physical monuments that celebrate people, the iconic statues representing previous masters, for example the felling of Saddam Hussein's statue witnessed globally, or the removal of communist heroes in post-soviet Europe. This destructive process reveals the symbolic weight of such statues and their significance as representational statements. Historically, they will generally have a strong link with the people of the territory where they are placed, rather than outsiders or foreign visitors. However, because of this very fact, they will become markers of distinction, cultural signifiers and therefore assets in the drive to differentiate one place from another in the global tourism market. The role of the statue and the public portrait is expanding as they become potential tourist attractions.

The awareness among policy makers and the general public of the economic importance of tourism and the part that cultural heritage can play in attracting visitors has led to an increase in the products of heritage (see Hewison, 1987) and promotion of heritage not only by policy makers, but also indigenous populations: this is the case in Dumfries and Galloway. Whereas, by contrast, for the Brito family of Bayahibe, Dominican Republic, action leading to the public recognition of local heritage is a statement of ownership and identity. This difference in the utilisation of heritage indicates different levels of power between population groups: in Dumfries and Galloway the residents have legal security over their property but desire to represent their own heritage and compete for additional outside resources, especially in the form of visitors; but, by contrast, in Bayahibe the villagers need to prove their legal right to their land first. Their ownership of the land, as the cohesion of the community in the sense of physical attachment to land and buildings, is continually threatened by powerful people in the shape of local and national governing bodies and big business interests. Primarily, in Bayahibe, the people need to make a successful claim to their cultural and natural heritage. They are currently in the process of creating their history and consolidating their rights to property. After this has been secured, they may utilise these assets to attract visitors. At the same time, powerful groups, including government agencies, are using the fishing heritage to publicise and market the region.

The recent increase in importance of cultural heritage as an attraction for tourists is due to factors including a bigger market overall, a wealthier and more educated tourist base, a changing demographic profile of tourist, a desire to offer something more than sunshine, sand and sea for competitive purposes. Consequently, cultural heritage becomes an additional attraction and a distinguishing element in a destination's inventory of products. This increase in potential value means that those who can gain from promoting aspects of culture will do so, and may be intent to display or interpret facets of their own culture that do not reflect well or fairly on all inhabitants, or simply give a false image.

Powerful groups have always been able to shape official historic records and public memory to their liking; however, with the introduction of a potential tourism market in mind, they will be more inclined to think of the public image abroad in order to attract visitors. Add into this mix the increasing popularity of constructing grass-roots heritage memorials and collections, or attractions based on indigenous culture, together with the ostensible democratisation and opening up of communication channels afforded by the internet, the relationship between power, tourism and cultural heritage is becoming increasingly complex and relevant.

References

Adams, R.N. (1977) Power in human societies: A synthesis. In R.D. Fogelson and R.N. Adams (eds) *The Anthropology of Power: Ethnographic Studies from Asia, Oceana and the New World* (pp. 387–410). New York: Academic Press.

Bianchi, R. (2004) Heritage, Identity and the Politics of Commemoration on "Columbus Island" (La Gomera, Canary Islands). *Working Papers in Tourism and Culture*. Sheffield Hallam University, Centre for Tourism and Cultural Change.

Boissevain, J. and Selwyn, T. (eds) (2004) *Contesting the Foreshore: Tourism, Society, and Politics on the Coast*. Amsterdam: Amsterdam University Press.

Bourdieu, P. (1986) *Distinction: A Social Critique of the Judgement of Taste*. London: Routledge.

Browne, S. (1994) Heritage in Ireland's tourism policy. In J.M. Fladmark (ed.) *Cultural Tourism: Papers Presented at the Robert Gordon University Heritage Convention* (pp. 13–26). London: Donhead Publishing Ltd.

Castellano-Gil, J.M. and Macias-Martin, F. (2002) *History of the Canary Islands*. Tenerife: Centro de la Cultura Popular Canaria.

Concepcion, J. (1989) *The Guanches Survivors and their Descendants*. La Laguna, Tenerife: La Cuesta.

DGTB (2001) *Dumfries and Galloway Area Tourism Strategy 2001–2006*. Dumfries: Dumfries and Galloway Tourist Board.

DGATP (2007) *A Tourism Strategy for Growth: 2007–2009*. Dumfries: Dumfries and Galloway Area Tourism Partnership.

D&G (2002) *Dumfries and Galloway Edition 2002*. Dumfries: Dumfries and Galloway Tourist Board.

D&G (2004) *Dumfries and Galloway: Where to Stay 2004*. Dumfries: Dumfries and Galloway Tourist Board.

D&G (2006) *Dumfries and Galloway: See and Do 2006*. Dumfries: Dumfries and Galloway Tourist Board.

Dobal, C. (1997) Herencia Espanola en la cultura Dominicana hoy. In B. Vega (ed.) *Ensayos Sobre Cultura Dominicana*. Santa Domingo: Museo de Hombre Dominicana.

Fernandez-Armesto, F. (1991) *Columbus*. Oxford: Oxford University Press.

Fladmark, J.M. (ed.) (1994) *Cultural Tourism: Papers Presented at the ROBERT Gordon University Heritage Convention*. London: Donhead Publishing Ltd.

Galvan-Tudela, A. (1995) *Los Procesos Etnicos en Regions Insulares: A Proposito de las Islas Canarias*. Utrecht: Isor-Press, Utrecht University.

Geertz, C. (1973) *The Interpretation of Cultures*. London: Fontana Press.

Greenwood, D. (1989) Culture by the pound: An anthropological perspective on tourism as cultural commodification. In V. Smith (ed.) *Hosts and Guests: The Anthropology of Tourism* (pp. 171–187). Philadelphia, PA: Pennsylvania University Press.

Hernandez-Hernandez, P. (1986) *Natura y Cultura de las Islas Canarias*. Tenerife: Litografia P.H.G. Romera S.A.

Hewison, R. (1987) *The Heritage Industry: Britain in a Climate of Decline*. London: Methuen.

Howard, D. (1999) *Dominican Republic: A Guide to the People, Politics and Culture*. New York: Latin American Bureau.

Hubbard, P. and Lilley, K. (2000) Selling the past: Heritage-tourism and place identity in Stratford-upon-Avon. *Geography* 85 (3), 221–232.

Keesing, R. (1981) *Cultural Anthropology: A Contemporary Perspective*. Orlando, FL: Harcourt Brace Jovanovich.

Latzel, M. and Reiter, J. (1998) *Insight Compact Guide: Dominican Republic*. Singapore: APA Publications.

Lefebvre, H. (1991) *The Production of Space*. Oxford: Blackwell.

Lowenthal, D. (1998) *The Heritage Crusade and the Spoils of History*. Cambridge: Cambridge University Press.

Macleod, D.V.L. (1998) Office politics: Power in the London salesroom. *Journal of the Anthropological Society Oxford* 29 (3), 213–229.

Macleod, D.V.L. (2002) Disappearing culture? Globalisation and a Canary Island fishing community. *History and Anthropology* 13 (1), 53–67.

Macleod, D.V.L. (2004a) *Tourism, Globalisation and Cultural Change: An Island Community Perspective*. Clevedon: Channel View Publications.

Macleod, D.V.L. (2004b) Selling space: Power and resource allocation in a Caribbean coastal community. In J.C. Carrier (ed.) *Confronting Environments: Local Understandings in a Globalizing World* (pp. 31–48). New York: Altamira.

Macleod, D.V.L. (2005) Narratives of belonging and identity in the Dominican Republic. In J. Besson and K. Fog Owig (eds) *Caribbean Narratives of Belonging: Fields of Relations, Sites of Belonging* (pp. 97–114). Oxford: Macmillan.

Macleod, D.V.L. (2006) Cultural commodification and tourism: A very special relationship. *Tourism, Culture & Communication* 6, 71–84.

McCrone, D., Morris, A. and Kiely, R. (1995) *Scotland the Brand: The Making of Scottish Heritage*. Edinburgh: Polygon.

Pascual, J.J. (2004) Littoral fishermen, aquaculture, and tourism in the Canary Islands: Attitudes and economic strategies. In J. Boissevain and T. Selwyn (eds) *Contesting the Foreshore: Tourism, Society, and Politics on the Coast* (pp. 61–82). Amsterdam: Amsterdam University Press.

Pons, F.M. (1997) Modernización y Cambios en La República Dominicana. In B. Vega (ed.) *Ensayos Sobre Cultura Dominicana* (pp. 211–245). Santa Domingo: Museo del Hombre Dominicana.

Robertson, J.F. (1994) *Story of Galloway.* Glasgow: Lange Syne.

Timothy, D.J. and Boyd, S.W. (2003) *Heritage Tourism.* Harlow: Prentice Hall.

TNS (2006) Scottish Recreation Survey: Annual summary report 2004/05. *Scottish Natural Heritage Commissioned Report* Number 183 (ROAME NO. FO2AA614/3).

Tylor, E.B. (1871) *Primitive Society: Research into the Development of Mythology, Philosophy, Religion, Language, Art and Culture.* London: John Murray.

Urry, J. (1990) *The Tourist Gaze: Leisure and Travel in Contemporary Societies.* London: Sage.

Vega, B. (ed.) (1997) *Ensayos Sobre Cultura Dominicana.* Santo Domingo: Museo del Hombre Dominicana.

VisitScotland (2006) *Tourism in Dumfries and Galloway 2005.* On WWW at http://www.scotexchange.net/dumfries-galloway-2005.

Chapter 5

Cultural Perspectives on Tourism and Terrorism

MICHAEL HITCHCOCK and I NYOMAN DARMA PUTRA

Introduction

Security is an issue that is important to tourism, thus stakeholders in this sector are heavily involved in the struggle for resources. Security as a state of affairs enables the utilisation of assets or resources such as beaches and cultural heritage, which attract tourists. It is a widely held view among tourism analysts that international visitors are very concerned about their personal safety (Edgel, 1990: 119) and that '... tourism can only thrive under peaceful conditions' (Pizam & Mansfield, 1996: 2). The reason is that there are many tourism destinations, such as Jamaica in the 1970s, which have acquired a reputation for being insecure and have thus seen their tourism arrivals crumble. A common way of making areas more secure and to allay fears of insecurity is for local and national governments to provide support for more intensive policing in tourism areas. This chapter concerns one aspect of security, terror, which can usefully be seen as a form of insecurity that leads to extreme instability, and it makes the point that anthropologists are well placed to shed light on such issues as security and terror, though these concerns are often seen to be the preserve of specialists on politics and security studies. With their emphasis on fieldwork and their interest in the cultures and contexts of the places in which tourism takes place, anthropologists can offer an especially nuanced understanding of security issues: the central case study of Bali in this chapter illustrates this point.

Perspectives on tourism and terrorism

Before considering what anthropologists can bring to the table, it is worthwhile mentioning the perspectives offered by other specialists concerned with security issues. Within the field of security and strategic studies, for example, there does not appear to be a substantial literature relating to tourism, but what seems to have caught the attention of analysts in this area is tourism's vulnerability to terrorism, especially with regard to the financial and political value of hostage taking. Analysts such as Rabasa for instance, have investigated the substantial sums procured by militant groups such as Abu Sayyaf, which took 21 hostages, including 10 foreign tourists, from a diving resort in the Malaysian state of Sabah. The kidnapping earned the militant Islamic group US$20 million, which was reportedly paid by Libya (Rabasa, 2003: 54). Security analysts such as Rabasa tend to stress the political under-pinnings of the material they analyse, which in his case relates to Southeast Asia, and concerns attempts by terrorists to remove modern borders of Southeast Asia to create a substantial Muslim Caliphate (Rabasa, 2003). Unsurprisingly, the governments of the region, including the country with the world's largest Muslim population, Indonesia, steadfastly oppose this agenda. Terrorism networks with local agendas that converge with those of al-Qaeda have surfaced with the arrests in Malaysia, Singapore and Indonesia of militants associated with *Jemaah Islamiyah* (JI); thus Southeast Asia has emerged as a major terrorist target and this has major implications for the region's important tourism industry.

There is also a growing body of literature from the perspective of tourism management or tourism studies, which understandably focuses on the implications of terrorism for developing and maintaining this important economic sector. These analysts largely concur that interna-tional visitors are very concerned with their personal safety (Edgel, 1990: 119) and that '...tourism can only thrive under peaceful conditions' (Pizam & Mansfield, 1996: 2). Political stability is considered to be a prerequisite for developing and maintaining a successful tourism industry, though it is recognised that this sector is vulnerable to international threats such as terrorism (Richter & Waugh, 1986: 238); analysts accept that it may be impossible to isolate tourists completely from the effects of international turbulence (Hall & O'Sullivan, 1996: 120). Security and peace may be crucial for tourism and international travel, but national and supranational organisations concerned with tourism have little influence on peace and security agendas (Hall *et al.*, 2004).

Many tourism promotion boards also share the perception that tourism is sensitive to crises and it is widely held that the media has a particular responsibility to help alleviate the fears of travellers. In this respect, the media is seen as being a major force in the creation of images of safety and political stability in destination regions (Hall & O'Sullivan,

1996: 107). It is not, however, direct threats to tourism, such as terrorism, crime and infectious diseases, that are seen as a cause for alarm, but negative reporting in general. For example, following the onset of the Asian monetary crisis in 1997, Thailand became so alarmed about the future of its tourism industry in the wake of the poor publicity that it sought to counter the flood of bad news by the positive promotion of the country as a cost-effective destination (Higham, 2000: 133). Thailand's positive promotion of tourism at a time of crisis is widely hailed as a success story, and the country has remained very sensitive about its image as a destination ever since.

Another group of authors, who might be considered to be analysts, though not in the social science sense, are novelists. Two in particular have explored the juxtaposition of tourism and terrorism within the context of the developing world: Giles Foden (2002) in *Zanzibar* and Michel Houellebecq (2002) in *Platform*. The former concerns the 1998 embassy bombings in East Africa. Foden's book follows in the long tradition of international thrillers combining several storylines from some of the genre's usual suspects. Foden's heroes are Nick Karolides, an American marine biologist who joins a USAID mission in Zanzibar, and Miranda Powers, an executive assistant of security at the US Embassy in Dar es Salaam. The two newcomers to East Africa become romantically involved and fall into the terrorists' web in the island tourism destination. Through the fictional character of Khaled, Foden provides some cultural insights into the world view of terrorists, in this case an African Muslim manipulated by outside fundamentalist forces. Khaled is based on the real-life Zanzibari terrorist, Khalfan Kamis Mohammed, who was sentenced in New York to life imprisonment in 2001 for his role in the 1998 bombing.

Houellebecq has probably received more critical attention than Foden, partly because of his provocative style. In his controversial novel, he combined an account of sex tourism with a horrific terrorist attack in Thailand and whatever the merits of the book, which was originally published in French in 1999, the author has the foresight to show how a tourist resort could become a terrorist target in Southeast Asia. Houellebecq's work may be fiction and, according to his critics, he is very opinionated, but this does not detract from one of the main messages of the book: tourists are easily attacked and some of the things that they engage in (e.g. sex tourism) may be used as a justification for attacking them.

Anthropological analyses of tourism and terrorism

Anthropologists have also been part of this debate and it comes as no surprise that cultural issues feature prominently in their analyses. One of the first, if not the first, of the anthropologists to consider this issue was Heba Aziz in a much-cited paper concerning attacks on tourists in Egypt. The paper notes that tourists had become targets to advance certain religious and political causes since the early 1990s at least, and that according to the Ministry of the Interior, terrorists had killed 13 tourists, as well as 125 members of the Egyptian security forces, since 1991 (Aziz, 1995: 91). The paper examined the relationship between Islam, hospitality and the concept of tourism and pointed out that Islam does not reject tourism *per se*. It was, however, the nature of tourism development, especially in Upper Egypt, that provoked acts of violence. The paper questioned whether tourists were the real targets of these attacks and argued that it was the tourism industry, the government and the developers, as well as the tourists, who were responsible for this undesirable situation. The paper, moreover, argues that the violence was a reaction to irresponsible tourism development and that clamping down on Muslim activism was merely damage limitation and not a solution, and that terrorism and xenophobia in Egypt were an indicator of a problem rather than a problem in its own right (Aziz, 1995: 95).

The publication date (1995) of Aziz's paper is significant, since it came out well before the widely publicised attack in Luxor in 1997, which left 58 foreign visitors dead. It was probably the scale of the attack in Luxor that started to move terrorism up the league of security worries concerned with travel and it is worth noting that it remains a top priority for Western governments. For example, the Australian Government's Department of Foreign Affairs and Trade on 5 January 2008 issued the following words of caution at the start of its summary of travel advice for Egypt:

> We advise you to exercise a high degree of caution in Egypt because of the high threat of terrorist attack. Attacks could occur at any time, anywhere in Egypt. (www.smartraveller.gov.au)

The Australian Government voices similar concerns and gives a high priority to terrorism in its travel advice for Indonesia, Malaysia, the Philippines, Thailand, Singapore, Sri Lanka, India and Pakistan. By contrast, travellers are warned in Vietnam to be aware of crime and in Nepal about the poor security situation.

Another anthropologist to have written a substantive analysis of a terrorist attack on tourists is Sally Ann Ness. She has likened the attacks in Bali to a terrorist incident at Pearl Farm Beach on Samal Island in the Philippines, which she sets apart from more economically related incidents of tourism-related violence that have occurred in the Philippines and elsewhere. She also notes the 'family resemblance' of the Pearl Farm assault to the Marcos-era outbreak of arson attacks on luxury hotels in the 1980s by the politically motivated Light-a-Fire Movement (as the activists came to be known) (Ness, 2005: 119). This movement, at times, combined economic motives with political ones, as could well have been the case with the Pearl Farm attack, but the motives of the Light-a-Fire Movement were not only concerned with generating revenue for dissident groups. Ness makes the point that the Pearl Farm attack was more closely linked to economically related forms of violence on tourism than with other forms, such as banditry (Ness, 2005). This kind of violence may be understood as a form of locational violence directed against a particular kind of place and not a particular person or collection of individuals. Ness argues that tourism landscapes with their consumption-oriented treatment of pre-existing cultural places can create a kind of disorientation that invites locational violence, but as this chapter argues, the bombings in Bali were concerned as much with place as with certain kinds of people.

The Bali Bombings

One of the most sustained investigations into terrorist attacks on tourists has been undertaken by the authors of this chapter with regard to the bombings in Bali in 2002 and 2005. The first part of the research was undertaken with funding from the British Academy in two main blocks of about a month's duration in July 2003 and July 2004, though work continued intermittently in between. Work resumed in 2006, following the second round of bombings and because of the sensitivity of the issues explored, all the informants remain anonymous. The research on the formal sector comprised semi-structured interviews ($n = 30$), usually in the informant's office, with Balinese tourism officials, the leaders of Indonesian trade associations and general managers of hotels. The authors also attended workshops designed to help the recovery of Bali's tourism industry, which were attended by representatives of local government, industry, NGOs, academia and the media whom the Minister of Culture and Tourism has characterised as the main stakeholders in Bali's tourism industry (Ardika, 2000). The authors also had access to the trial transcripts of the alleged bombers and studied the website reports of journalists who claimed to have interviewed the bombers.

The 2002 Bali bombings

The bulk of the research was devoted to the three bombings on the night of 12 October 2002, which resulted in the largest death toll ever recorded from an attack on a holiday resort. The three targets were the Sari nightclub and Paddy's Bar in Kuta and the American Consulate in Denpasar. The bomb at Paddy's Bar did not at first appear to have had a great impact, but it had a deadly side effect. It drew people on to the streets so that when the next bomb at the nearby Sari Club went off, more people were exposed. The explosives had been packed into a van that had been parked outside the packed nightclub, which was almost entirely destroyed by the blast and the raging fire that ensued.

The majority of those killed were Australians (88) and the second largest loss of life was borne by Indonesia (35), the majority of whom were Balinese and, moreover, Balinese Muslims. Bali is famously a majority Hindu island, but it has a small Muslim community, many of whom are ethnically Balinese. The third largest death toll was suffered by the UK (23) and many other victims were of European origin: German (6), Swedish (5), French (4), Danish (3), Dutch (4), Swiss (3), Greek (1), Portuguese (1), Italian (1) and Polish (1). There were victims from the Americas – USA (7), Canada (2), Ecuador (1), Brazil (2) – as well as elsewhere in Asia: Japan (2), Taiwan (1), Korea (2).

Led by the police inspector, General I Made Mangku Pastika, the alleged perpetrators were apprehended quite quickly following appeals to the population at large to be vigilant and to report anything suspicious. The bombers had covered their tracks, but a slip-up alerted Pastika to the first of the suspects, Amrozi bin H. Nurhasyim. Amrozi was a mechanic who had hoped to confound the police by altering the registration number on the van used to transport the larger of the bombs to the Sari Club. What Amrozi did not know, however, was that the van had been in service as a minibus and thus bore another number that he had failed to notice, which Pastika's investigators were able to trace. After his arrest, Amrozi gave the police a detailed confession, which underpinned much of the investigation that ensued.

Unlike many other terrorist attacks in which the bombers perish or are kept away from the press and public when apprehended, it was possible to examine the motives of the Bali bombers with some clarity. This was because their trials were held in public in front of a phalanx of the international media, some of whom, at least in the early days, were able to personally interview the bombers. Initially, the bombers claimed that their intended victims were Americans: but when it dawned on them that the largest numbers killed were of Australian origin, and the second largest, Indonesian, they modified their position. They offered a variety of political justifications for their attacks and their actions appeared to be supported by a statement, allegedly from Osama bin Laden, that it was indeed Australians who were being targeted because of their alliance with the USA. Comparatively few victims may have been Americans, but what should not be lost sight of is the fact that the USA was a target, if not the main target, since its consulate was also bombed that night, though there were no casualties, whereas the Australian consulate was not targeted.

Coupled with the confusion surrounding the precise nationality of the intended victims was the reaction of the press to the behaviour of the alleged bombers, notably Amrozi and Mukhlas. The latter was delighted when his death sentence was read out a year after the bombings, much to the astonishment of representatives of the world's media gathered in and around the courthouse in Bali's capital, Denpasar. Along with his younger brother and the operation's mastermind, Imam Samudra, Mukhlas was the third bomber to be sentenced to death for his role in the attack. Mukhlas' response echoed that of his brother, Amrozi, nicknamed the 'smiling bomber' by the press, who had been sentenced earlier, claiming that there were many in Indonesia willing to take his place should he die.

Once charged, Mukhlas and his henchmen did not deny that they had been the perpetrators, even to the extent of correcting the judges to ensure that the record was straight. The only exception was Ali Imrom, who confessed that the attacks had been against his Muslim teachings. Ali Imron also wore a suit and tie in court, behaving politely and expressing remorse, even weeping a couple of times in public. By contrast, the other bombers donned Muslim-style clothes, including the Indonesian fez (*peci*) and sarongs, during the trial and were photo-graphed carrying out their devotions.

The bombers expressed pride in their achievements and even claimed to be unconcerned about the deaths of their fellow citizens despite the fact that many of them were Muslims. Amrozi simply offered to pray for the dead Balinese, but appeared to be unshaken in the belief that he had done something worthwhile; expressing regret that he had not killed more Americans. Amrozi was candid about his motives, claiming that he had learned about the decadent behaviour of white people in Kuta Beach from Australians, notably his boss in Malaysia, becoming incensed about their stories of drug-taking and womanising. The Malaysian connection was important in a technical sense since he had worked alongside French and Australian expatriates in a quarry and had learned about explosives, as well as about how easy it was to attack Bali.

Amrozi's antipathy toward Westerners may have been nurtured by his experiences in Malaysia, but his exposure to Islamic radicalism could also have been a factor. He attended, for example, the Lukman Nul Hakim College in the 1990s in Malaysia, where Abubakar Ba'asyir taught, though it remains unclear what he studied. By 1996, Amrozi had become convinced that it was the Jews who sponsored Westerners and that they were intent on controlling Indonesia. As his hatred developed, he became convinced that violence was the only way to get Westerners out of Indonesia; diplomatic means had, for him, proved ineffectual.

In addition to Westerners, Amrozi appears to have equally disliked non-Muslim Indonesians and claimed to have been involved in attacks in Jakarta, the Indonesian capital, and in the religiously divided island of Ambon. Amrozi also maintained that he had participated in the Christmas Eve attack in 2000 in Mojokerto, Central Java, which claimed 19 victims and admitted that he had had a hand in the attack on the Philippines Embassy in Jakarta and had actually mixed the explosives. Likewise, Imam Samudra, who is known by other aliases and had access to higher education, seems to have intensely disliked non-Muslims and had honed his bomb-making skills in Afghanistan. Imam Samudra was also suspected of being involved in a series of church bombings across Indonesia. In his evidence at a separate trial, that of Abubakar Ba'asyir, Imam Samudra claimed that the bombings were part of a *jihad*, though he denied any links with the militant group, *Jemaah Islamiah*. When asked about the Christians who died in those attacks, he responded by saying that 'Christians are not my brothers'.

What is significant is the bombers' self-perception as not being terrorists, but legitimate combatants, especially Imam Samudra (2004), who clearly saw himself as a holy warrior and is in fact the author of a 280-page book, entitled *Aku Melawan Teroris* (I Oppose Terrorism), which he wrote in prison. He legitimised his attacks by citing the Koran and reaffirmed that he was involved in holy war in Bali and that his target was the USA and its allies, which he called the 'nations of Dracula'. According to his version of Islam, these enemies could be killed wherever they could be located. Searching at random, he came across the Sari Club and Paddy's Pub, drawing the conclusion that they provided a perfect potential target (Samudra, 2004: 120). Imam Samudra also expressed concern about the large number of Indonesians who had died at his hands, writing that that it was 'human error' (English is used in the original), which he much regretted (Samudra, 2004: 121).

Significantly, what emerged from the trials in Denpasar is that tourists per se were not the intended victims, but Westerners and non-Muslims, especially American allies. Tourists were targeted because of their perceived association, no matter how remote, with attacks on Muslims and as Amrozi claimed, he felt no remorse in killing them: 'How can I feel sorry? I am very happy, because they attack Muslims and are inhumane' (*Asia Times*, 3 June 2003).

The bombers expected more of their victims to be Americans, but when informed that the majority of their victims were Australians, one of them quipped: 'Australians, Americans whatever – they are all white people' (*Asia Times*, 3 June 2003).

Amrozi's indifference to the suffering of innocent victims may be a reaction to the alleged abuses conducted by Western nations, but also seems to suggest that their whiteness was a factor as well. The manner in which a person is characterised by their physical attributes in Indonesian can be ambiguous and may range from the culturally neutral *orang putih*, literally 'white person', to the more controversial *bulé* (albino). In Indonesian usage, 'albino' can be neutral and often crops up in humour, but when applied dismissively as in the quotation above, it can convey notions of inferiority. In Sarah Ferguson's interview of 23 May 2003, Amrozi is translated as using the term 'whities', however, it remains unclear what was actually said in Indonesian. But what may be more significant is that the word for tourist, *turis*, which is often used to refer to white people only, does not appear to have been used, which suggests that it was the victims' Western or white attributes that influenced the bombers' choice of target.

Mukhlas, the eldest of the three brothers and a veteran of Afghanistan, claimed that the USA had not always been the target and that he had been involved in the struggle against the Soviet Union alongside Osama bin Laden.

> Osama bin Laden. Yes, I was in the same cave as him for several months. At the time, he wasn't thinking about attacking America. It was Russia at that time. (NineMSM, 23 May 2003)

In an English translation of an interview recorded in prison by Sarah Ferguson, he clearly explains the political reasons underpinning his attacks on Australians:

I want the Australians to understand why I attacked them. It wasn't because of their faults, it was because of their leaders' faults. Don't blame me, blame your leader, who is on Bush's side. Why? Because in Islam, there is a law of revenge. (NineMSN, 23 May 2003)

What appears to have united the bombers is their unshaken belief that they were participating in a *jihad*, a struggle to establish the law of God on earth, which is usually interpreted as meaning holy war. The term *jihad* has two meanings, the first being a struggle of any kind, especially a moral one, such as striving to be a better person, a better Muslim; the struggle against drugs, immorality and infidelity. Its second meaning refers to holy war, which is embarked on when the faith is threatened, but only with the approval of the appropriate religious authority (http://www.faculty.juniata.edu/tuten/islamic/glossary.html). The bombers do not appear to have had the necessary authority to carry out their attacks and, moreover, had not abided by the rules of *jihad*, which are quite specific, especially with regard to attacking without warning. Ali Imron, in particular, was aware of this and confessed in court (15 September 2003) that the bombers had broken the terms of *jihad* and that the bombings were wrong whatever their motives: '... whatever the motive behind the Bali bombings, the act was wrong because it breached the rules'.

There could well have been some confusion about the technicalities of *jihad*, since there was little discussion about its meaning in Indonesia at this time, but this changed after the second round of bombings on 1 October 2005. When videos of the suicide bombers' confessions recorded before the attacks were circulated, religious leaders felt compelled to comment, with the majority speaking out against the practice of suicide bombing. They also argued that the instigation of *jihad* was only acceptable when the nation was under attack and that, in contrast with Iraq, Indonesia was not threatened.

Security resources: Policing

Security was highly prioritised in the aftermath of the bombings and anything that detracted from the desired image of Bali as a safe haven was viewed with alarm. Inspector General I Made Mangku Pastika, the police chief who led the investigation that rounded up the bombers, was given a clear mandate from the island's governor to prioritise the safety of visitors. Pastika presided over a complete change in policing policy on the island in response to requests from businesses and the public alike, but he was hampered by a legacy of petty corruption and police inefficiency dating from the Suharto era. Even before the bombings took place, concerns had been raised about the image of the Indonesian police force; many hotels and shopping centres were unwilling to see uniformed police protecting their premises because they felt that it heightened the sense of insecurity. Before the bombings of 2002, the emphasis was on covert policing and reacting to criminal activity by deploying the police in public when they were needed. The police were often out of sight and confined to their stations, waiting to attend emergencies as and when they arose. As the island recovered from the bombings, visible policing came to be seen as a priority, spurring a debate about issues such as the appearance of men in uniform.

Well before the bombings, the tourism authorities had wanted more user-friendly police uniforms, and even dedicated tourism training for officers; they used the crisis to make the case for a larger allocation of resources. The first immediate result was the tripling of the number of intelligence officers, partly because of renewed fears concerning a fall-out from unrest in the restive Sumatran province of Aceh and raised apprehension about the possibility of additional terrorist attacks. The newly recruited officers were deployed at all points of entry, notably the international airport of Ngurah Rai and the harbour of Gilimanuk, the main entry port from Java, which had been used by Amrozi and his fellow conspirators. The second significant change was the inauguration of beach policing, *polisi pariwisata* (tourist police), on Kuta Beach in the 'Baywatch' style associated with the USA and Australia. Foreign expatriates in Bali also volunteered to help improve the spoken English of the new 'Baywatch' squad and to teach them Western manners. A short-term accreditation scheme was also created to ensure that police were aware of the layout of hotels in the event of emergencies, and to upgrade standards in general advisors were brought in from Japan and Australia.

The 2005 Bali bombings

Bali was bombed again on 1 October 2005 when cafes along Jimbaran Bay and Kuta were attacked, leaving 20 dead including three suicide bombers, most of whom were Indonesian citizens. The bombers may have killed fewer people, but to the alarm of the security forces, the bombs were more advanced and contained ball bearings, some of which were found in the bodies of injured victims. Within two days, the police had concluded that the bombings were the work of not only terrorists but also suicide bombers, after studying a video recorded by an Australian tourist who happened to be outside a restaurant targeted by the bombers shortly before the attack. The tourist accidentally recorded a man with a backpack, who was walking faster than the others around him, entering the restaurant. At the press conference that followed, General I Made Pangku Pastika showed journalists how a suicide bomber carrying a backpack could be seen walking among guests having dinner in the restaurant.

The police hoped that by circulating a poster with pictures of the three suicide bombers in colour, they would be able to identify the bombers, but several weeks passed without much of a response. To make them more identifiable, the pictures were cleaned up by removing the blood and debris on the bombers' faces, but once again there was little progress. It began to occur to the police that the bombers were possibly foreign nationals and that Indonesian citizens were simply unable, as opposed to unwilling, to identify them. The police and media began to speculate that the bombs were the work of two Malaysian fugitives from the Bali bombings of 2002, Azahari and Noordin M Top, and that a new generation of bombers had become involved.

Unlike the police enquiries of 2002, which had involved appeals to the public, the investigation of 2005 was more secretive with far less media coverage, possibly because of fears of a more global dimension to the latest attacks. The police only revealed that there were no significant developments, and that they were continuing to question witnesses, whose number rose above 700. In tandem with these enquiries, the police launched covert operations, shaking out alleged *Jemaah Islamiyah* suspects throughout Java, although no arrests were announced until after the storming of Azahari's safe house in Batu in Malang, East Java. Azahari and one of his followers were killed during the raid and the police found dozens of vest bombs, VCDs, books and a plan for a 'bomb party' for Christmas and New Year. After almost seven years on the run, Noordin was finally killed by Indonesian anti-terror police on a raid in a village of Central Java, 17 September 2009. His killing, which was followed by the killing of several of his followers, substantially reduced the capability of this terror group to carry out an attack.

According to Sydney Jones of the International Crisis Group, Noordin Top had started to refer to his splinter group as 'al-Qa'ida for the Malay archipelago', although he still regarded himself as the leader of JI's military wing. Jones also maintained that Noordin Top and the followers were wedded to the al-Qaeda tactic of attacking the USA and its allies wherever they could be located and since they were based in Indonesia, nearby Australia was a prime target (Radio Australia, *AM*, 6 May 2006). The resources needed to mount such attacks could have been provided by al-Qaeda, but are just as likely to come from the group's own criminal activities, an example being a raid on a gold shop in West Java shortly before the bombings of 2002. The proceeds helped to defray the expense of the attack, including an estimated Rp 3–4 million (around US 300–400) to make a vest bomb, the rental of premises and the costs of surveying the target area.

The details of the 1 October 2005 attacks appeared in papers found at the scenes of the bombings and in the hiding places of those taken into custody. The papers revealed how JI members surveyed potential targets in Bali and reported their findings to JI's master bomb-maker Azahari. They studied nightclubs, temples, shopping areas, sports venues, fast food outlets, souvenir shops and the airport, and concluded that Jimbaran Bay, the eventual scene of two attacks, was a good target. Moreover, they estimated that there would be at least 300 people there, thereby maximising the potential number of casualties (Wockner, 2006). One of the four suspects of the 2005 attack, Mohamad Cholily, said he was with Dr Azahari when they heard news of the bombings on BBC Radio. He claimed that Azahari had shouted, 'Allahu Akbar' (God is Greatest) and 'Our project was a success'. Cholily, who was learning bomb-making skills from the so-called 'demolition man' (Azahari), was arrested one month later. It was also Cholily who led police to the safe house in East Java where the famous fugitive was hiding (Wockner, 2006).

Azahari was killed in the raid, but this did not greatly alleviate the public's fears, largely because of the existence of a plan for a 'bomb party', in which many targets would be struck simultaneously. Even though the police confiscated numerous vest bombs, it was widely believed that Azahari must already have recruited dozens of people who were prepared to conduct suicide missions. Anxieties were also heightened by the video footage recovered in the operation because they contained the pre-recorded confessions of the three suicide bombers who attacked Bali: Salik Firdaus, Aip Hidayatulah and Misno.

The confessions were widely circulated in the media, both in Indonesia and abroad, and conveyed the terrifying message that further attacks were planned. The Australian government responded by issuing additional travel warnings, leading to a decline in visitor arrivals, but there were important differences as compared with the 2002 attacks. For example, the massive exodus of tourists that had followed the 2002 bombings did not re-occur and it looked at first as if the tourism industry would not be so adversely affected. Eventually, the numbers began to drop drastically, possibly due to the combination of the travel warnings and the televised confessions of the suicide bombers. Terrorism in its global context also appears to have exerted an influence, as Indonesians were shocked by coverage of a female Iraqi suicide bomber who succeeded in bombing a wedding party in Amman, including an Indonesian musician among her victims.

Security as a Resource

Bali had been one of the most peaceful Indonesian islands since the strife of the mid-1960s that accompanied the demise of Sukarno and it remained largely untroubled during the violent years that followed the Asian Crisis and the collapse of Suharto's longstanding regime in 1998. Bali's populous neighbour, Java, suffered greatly in the Asian Crisis with major riots and bombings and it is perhaps not surprising that Bali's comparative security and prosperity may have engendered a certain amount of envy. Many Balinese also appear to have been oblivious to potential threats within Indonesia, with many believing that theirs was a 'sacred island' protected by God. This perspective seems to have been reinforced by an incident in the 1980s when a bomb sent from Java exploded on a bus before it reached Bali. The fact that Bali did not suffer from the same kinds of problems that had engulfed other Indonesian islands may have led to a certain amount of complacency in the island's security services.

What Bali's security services do not seem to have been especially aware of was that the island's status as a major international tourism destination made it a tempting target, since any attack on it was likely to generate a high level of media interest, not least because of the presence of Western interests and Western tourists. Foreign-born tourists are also invaluable because there is less risk of a public backlash when they are attacked instead of 'innocent local victims', though this backfired in the case of the Bali bombing because of the high mortality rate among Indonesians. Another consideration was the fact that the deaths of foreign nationals would be likely to generate external publicity that the government could not suppress. Bali's global profile and prosperity may have attracted the attention of terrorists, but what seems to have been decisive was that it was above all an easy target. Another factor was the tightening of security elsewhere in Indonesia, not least in the capital Jakarta, which made other potentially newsworthy targets more difficult to attack. As the strife that accompanied Suharto's fall from office spread, security measures were tightened to protect embassies and government institutions. A tourism resort in Bali would have appeared comparatively easy to attack: tourists were often present in large numbers and were difficult to protect without curtailing their freedom; they also had the advantage of following predictable behaviour patterns and a tendency to cluster.

An attack on Bali's tourism industry would not only be likely to generate international media interest, but would also wreak economic damage as the ensuing publicity kept visitors away. This turned out to be the case as Bali remained in the media spotlight not only because of the bombings, but also because of the trials that followed. International visitor numbers fell from 1,285,844 in 2002 to 994,616 in 2003; Erawan (2003: 265) has argued that the bombings of 2002 had by far the biggest impact on Bali's economy of any recent crisis: in 2000 the tourism sector contributed 59.95% of provincial GDP, but in 2002 it had fallen to 47.42%. By 2007, international visitor numbers were down to 910,567, but in the first six months of 2008 they had improved on the previous year by 21% (source: ETN Global Travel Industry News).

Conclusion

Many national governments rate terrorism highly as a risk for travellers and list these concerns on their official websites. Likewise, many tourism studies specialists rightly place emphasis on the importance of security as a resource if a successful industry is to be developed and maintained, but a considerable amount of local knowledge is necessary to explain why this security is needed in the first place and show what might be done to help ameliorate such attacks. One of the first things to note is that terrorists are not uniform in their motives and may have very different reasons for attacking tourists and tourism facilities. Risk assessments that guide tourism policy and planning, as well as decisions on security, are often quite technical and market oriented, and ignore important cultural considerations. Aziz and Ness's work in Egypt and the Philippines, respectively, has shown, for example, that attacks on tourists by terrorists may have a connection with the way the tourism industry is developed, whereas the authors of this chapter have drawn very different conclusions in Bali. The complexity revealed by fieldwork and other anthropological investigative methods demonstrates the value of this type of research, which highlights the cultural context and utilises local knowledge in the pursuit of understanding.

The Indonesian bombers of 2002 offered different variations of the main reasons for their attacks, ranging from a simple desire to hit back at

Westerners for their supposed attacks on Muslims, to a more politically sophisticated attack on John Howard's support for President Bush and Australian intervention in East Timor in 1999. Some of their explanations have been couched in terms of what appears to be racial hatred, though these threats and statements are somewhat vague. What is clear is that they decided to bomb a tourist resort because it offered a relatively soft target: not because the victims were tourists per se, but because their numbers were likely to include large numbers of foreigners whose deaths would attract publicity to the terrorists' cause. Some disapproval over the alleged behaviour of tourists in Indonesia was expressed by the bombers, but it was the intended victims' nationality and perhaps racial type, their invaluable foreignness that appears to have been uppermost in the bombers' minds; tourists are useful because they create more publicity than when only locals are involved. Such publicity is moreover difficult to suppress, thereby enabling terrorists to make their various causes known more widely. Bali also seems to have been doubly attractive because any local victims would be likely to be Hindu and not Muslim, the faith of the attackers. As it happened, the bombers miscalculated and ended up killing significant numbers of their co-religionists.

After the 2005 Bali attack, police found a document called the 'Bali Project', which contained the reasons for targeting. The documents began with the question 'Why Bali?' to which the answer was: 'Because it is the attack that will have global impact. Bali is famous all over the world, even more famous than Indonesia. The attack in Bali will be covered by international media and the world will get the message that the attack is dedicated to America and its allies'. This turned out to be an accurate prediction since media worldwide covered the Bali attacks immediately.

The impact of the Bali bombings of 2005 on the island's tourism sector seems to be far worse than that of 2002. After the 2002 bombings, multinational investigations and support from the international community helped to speed up the investigation and restore Bali's image as a safe destination. Tourism arrivals recovered quite quickly once the island seemed secure again. But after the 2005 bombing, less help from the international community was evident due to a combination of factors: compassion fatigue in the aftermath of the tsunami, especially with Australia, which had contributed generously, and a stretching of resources in a generally less safe environment. Possibly because the 2005 attacks had a limited direct effect on Australians, less help with police work was offered to Indonesia; a wide range of considerations, including the identity of the victims, would appear to complicate the recovery of tourism from a terrorist attack.

The common feature of the attack in Thailand imagined by Houelle-becq and the real attacks in Bali is that they occurred in mass tourism resorts and that terrorists exploited the opportunities that this kind of tourism provides: relatively easy targets, large numbers of potential victims, relatively small numbers of co-religionists, the publicity value of foreigners and, finally, the alleged hedonism of tourists that could be exploited rhetorically as a justification for killing them. Interestingly, both Houellebecq and Foden provide some cultural insights into the perceptions and motives of terrorists, which is something anthropologists have been trying to do. An important point to note is that terrorists do not necessarily see themselves as such: indeed, one of the convicted bombers involved in the 2002 attack has written a book making precisely that point.

References

Ardika, I. Gde. (2000) *Seminar Sehari: Sekolah Tinggi Ilmu Ekonomi*. Jakarta: Kantor Menteri Negara Kebudayaan dan Pariwisata (Office of the Minister of State Culture and Tourism).

Aziz, H. (1995) Understanding attacks on tourists in Egypt. *Tourism Management* 16 (2), 91–95.

Edgell, D.L. (1990) *International Tourism Policy*. New York: Van Nostrand Reinhold.

Foden, G. (2002) *Zanzibar*. London: Faber and Faber.

Hall, C.M. and O'Sullivan, V. (1996) Tourism, political stability and violence. In A. Pizam and Y. Mansfield (eds) *Tourism, Crime and International Security* (pp. 105–121). New York: John Wiley.

Hall, C.M., Duval, D. and Timothy, D. (eds) (2004) *Safety and Security in Tourism: Relationships, Management and Marketing*. New York: Haworth Press.

Higham, J. (2000) Thailand: Prospects for a tourism-led economic recovery. In C.M. Hall and S. Page (eds) *Tourism in South and South-East Asia* (pp. 129–143). Oxford: Butterworth-Heinemann.

Houellebecq, M. (2002) *Platform*. London: Heinemann.

Ness, S.A. (2005) Tourism-terrorism: The landscaping of consumption and the darker side of place. *American Ethnologist* 32 (1), 118–140.

Pizam, A. and Mansfield, Y. (1996) Introduction. In A. Pizam and Y. Mansfield (eds) *Tourism, Crime and International Security* (pp. 1–17). New York: John Wiley.

Rabasa, A.M. (2003) *Political Islam in Southeast Asia: Moderates, Radicals and Terrorists*. Oxford: Oxford University Press for the International Institute for Strategic Studies.

Richter, L.K. and Waugh, W.L. (1986) Terrorism and tourism as logical companions. *Tourism Management* 7 (4), 230–238.

Samudra, I. (2004) *Aku Melawan Teroris* (I Oppose Terrorism). Solo: Jazera.

Wockner, C. (2006) How Bali bombers planned mission. *The Advertiser*, 29 April.

Part 2

Tourism and Culture: Presentation, Promotion and the Manipulation of Image

JAMES G. CARRIER

The four chapters in this section look at the relationship between tourism and one of the core concerns of anthropology. That concern is meaning, the significance that things have, manifested in images of them. In doing so, these chapters help illuminate some of the less obvious aspects and consequences of tourism of different sorts for the host countries in which it occurs, and the less obvious relationships between tourism and the social and political processes at work in those countries.

The sort of meanings and images that these chapters consider is that associated with groups of people and their practices, ranging from an entire country at one extreme, to specific and relatively small groups at the other. This sort of meaning is especially salient in tourism in poorer countries, the focus of these chapters. Some such countries continue to rely on the conventional attractions of sun, sand and sea in their coastal areas; and Chapter 9 is about one of these countries, Jamaica. However, such tourism is highly competitive, for one tropical beach is, after all, very much like another. Consequently, it is decreasingly attractive to countries that see tourism as a vehicle for economic growth. Many of these countries have, instead, been promoting the distinctive cultural groups within their borders that are exotic and attractive to visitors. The images and meanings associated with such groups, aimed outward at tourists, can be significant when seen in terms of the social and political context of the host country.

Chapter 6 presents a clear, straightforward example of this sort of cultural tourism, its images and meanings. In 'Tourists and Indigenous Culture as Resources: Lessons from the Embera Cultural Tourism in Panama', Dimitrios Theodossopoulos describes the Embera of Parara Paru on the Rio Chagres. Parara Paru is close enough to Panama City and the people there are exotic enough that it has been able to attract a substantial number of tourists, who travel up the river to the village and get several hours of Embera music, song, dance and food before travelling back late in the day. Theodossopoulos looks at two, intertwined sets of meanings that deserve consideration because of the recent boom in Embera cultural tourism.

The first of these is what it means to be Embera, and as Theodossopoulos notes, cultural tourism has a significant influence on this. Panama has had a long history of denigrating *Indios*, treated as backward undesirables. However, the government has recently encouraged tourism, and especially cultural tourism of the sort that the Embera practice. As a consequence of being attractively exotic, they have ceased to be mere *Indios* and have instead become a symbol of the country's diversity displayed in brochures and websites. Thus, the spread of cultural tourism has not just benefited the Embera economically. It has also benefited them politically, increasing their status within the country, and so increasing their power relative to the government and other groups in the country.

However, their very success as a tourist attraction has raised the second question that this chapter addresses: the meanings of their culture. Cultural tourists want to see authentic cultural forms, the truly exotic. At the same time, however, many critics argue that cultural tourism produces only a debased local culture, one that is shaped by the demands of tourism rather than local social and cultural processes. Theodossopoulos argues that tourism has not led to a debasement, but to a strengthening of culture in Parara Paru. As he notes, the Embera in the area have long survived by trade along the river; but instead of shipping loads of plantains down to urban markets as they did previously, they now ship loads of tourists up to their village. More importantly, it is they who control their dealings with tourists, not an outside company or entrepreneur. As a result, these Embera can control what is presented to tourists and what is not, who presents it and how, and the distribution of the income that it generates. Further, their touristic presentations repeat and strengthen important elements of their culture, and encourage children to learn them and stay in their village, rather than move elsewhere in search of a livelihood.

The case of the Embera shows the benign side of cultural tourism. For those living in Parara Paru, it has lead to increased material well-being and greater legitimacy and power within Panama, as well as what appears to be a resurgence of interest in indigenous practices. However, the Embera living further from Panama City, less accessible to tourists, derive fewer advantages. They benefit from the increasing relative status of *Indios*, but the main economic benefit that they derive is from being manufacturers of Embera artefacts that are shipped down to Parara Paru and sold to tourists there. While this inter-Embera trade is amicable, it increases the economic dependence of more distant villages on Parara Paru and so increases the power of that village over them. Cultural tourism, then, has different effects in different Embera villages, and it affects the relationships between them. This alerts us to the point that groups like the Embera, which may be attractive to tourists, are not necessarily uniform entities. Rather, different Embera live in different places, and tourism affects them differently.

Chapter 6 is concerned primarily with how cultural tourism affects the specific exotic group that is a tourist attraction, the Embera of Parara Paru. The next two chapters are oriented rather differently, not toward the specific groups that attract tourists, but toward the countries in which this cultural tourism occurs. Because that tourism revolves around sets of people who are presented as distinctive, it highlights difference rather than uniformity. The tourists travelling up the Chagres, after all, were not seeking what was common to Panama; they were seeking what was distinctive to the Embera. This stress on difference and distinctiveness has a special significance in countries where national identity itself is uncertain and disputed, for that tourism can both reflect and shape the understandings of and tensions among different segments of a country's population.

Chapter 8, 'Tourism and the Making of Ethnic Citizenship in Belize' by Teresa Holmes, describes a relatively new nation. Formerly British Honduras, Belize became independent in 1981 and faced the problem of welding its population into a nation united by something other than their colonial past. One important way that the government sought to address this political problem was through the medium of meanings, particularly what it meant to be Belizean. It sought to strengthen national identity amongst its citizens with the progressive model of the melting pot, which recognised the diversity of the country but located it in the past. The Spanish-speaking Central Americans, Amerindians descended from the ancient Maya, Garifuna or Black Caribs, the Creole groups, all existed, but were being transcended by the development of a common Belizean identity, and a common orientation to the nation-state and the government that represented it.

However, late in the 20th century, Belize placed increased importance on tourism in its economic policies. Initially, this was ecotourism, which had no obvious importance for the government's efforts to produce a common national identity. However, in an attempt to increase tourism revenues, government began to focus on cultural tourism, which stressed the very differences that hitherto it was trying to transcend. In this circumstance, the older project of nation-building had to change: the melting pot gave way to multiculturalism, signalled by the abandonment of plans for a National Museum, in favour of four regional Houses of Culture. As a consequence, what it meant to be a Belizean began to change. The older stress on national identity was replaced by a stress on being an exemplar of one's group within Belize, and on an appreciation of the distinctive virtues of one's own and other groups.

This shift in economic strategy reflected and sought to take advantage of changes in the tourism market. However, as Holmes notes, it also reflected and sought to accommodate changes in the country's population, and especially their power to define what it meant to be Belizean. Immigration from nearby Central American countries increased the number of Spanish speakers and Mayans in Belize, a country that was predominantly English speaking. Their efforts to secure their legitimacy in the country led them to challenge the image of the national citizen that the government had been supporting; their increasing numbers made those efforts harder to resist, as it made the shift from a policy of national unification to one of tetrapartite multiculturalism more attractive.

The orientation of Holmes's description of Belize complements that of Theodossopoulos's description of the Embera village of Parara Paru. Theodossopoulos was concerned primarily with the ways that the increasing importance of cultural tourism affected the Embera and groups like them, attractive and relatively accessible to tourists. On the other hand, Holmes's more national perspective shows how changing fashions in tourism intersected with changes that were taking place in national society. In Belize, this combination appears to undercut efforts to create a relatively unified nation, thereby reducing the authority of any government that seeks to speak for the country as a whole.

Where Holmes's chapter describes the effects of cultural tourism on what it means to be a citizen of a new country, Chapter 7, 'On 'Black Culture' and 'Black Bodies': State Discourses, Tourism and Public Policies in Salvador da Bahia, Brazil' by Elena Calvo-González and Luciana Duccini, describes those effects on what Blackness means in a state and city in a country that is well established. In doing so, it helps show how the concern with particularity that is part of cultural tourism can be linked with social and political cleavages within the host country. In Brazil, the state of Bahia and its city of Salvador have long been seen as especially Black, and historically, as Calvo-González and Duccini describe, Blackness has been seen in ways that oscillate between being cultural and being racial.

Since the Second World War, Brazilian policy increasingly resembled the melting pot that was the aspiration of the Belize government early in the 1990s. Just as German immigrants were seen as being transformed into Brazilian Germans by their history in the country, so Blacks were seen as being transformed into Brazilian Blacks. However, as was the case with Belize, in Brazil the government sought to expand tourism by identifying distinguishing cultural and historical characteristics for the country's different regions and cities. For the state of Bahia, this meant Blackness, and Salvador became the Roma Negra. One result was a change in what Blackness means, as distinctive ancestral roots gained importance relative to common Brazilian experience.

The increasing importance of African roots appeared in many forms, but Calvo-González and Duccini focus especially on its relationship with Candomblé, a religion that is exotic and attractive to tourists, which, increasingly, has been presented as African and is taken to characterise Salvador's Black population. Seeing Blacks as adherents of Candomblé and seeing Candomblé as being African meant seeing Blacks as African, fairly untouched by their centuries in Brazil. This facilitated a shift to understandings of Blackness that were more racial, more allied to an inherited biological state than to historical experience. This under-standing appears in the government's Black Population Health Group, which encourages attention to medical conditions taken to be character-istically Black and which is associated with Candomblé groups.

In tracing this shifting understanding of Blackness in Salvador, Calvo-González and Duccini make more insistent the question raised by Holmes's discussion of Belize: what are the systemic effects of a form of tourism that stresses difference? Holmes addressed this question by looking at the nation and its citizens, at events and processes at the national level and the ways that they were associated with individual identity. Calvo-González and Duccini extend this by looking at how the meaning of Blackness shapes and reflects the legitimacy and power of different groups within Salvador, and the ways that different people are or are not assigned to those groups. For instance, they begin their chapter by noting that the association of Blackness with Candomblé by both tourism and the state government excludes the high proportion of Salvador Blacks who adhere to other religious beliefs. Later in their chapter, they note how the concern to encourage what is seen as traditional street-vending of local foodstuffs excludes those who simply want to sell those foodstuffs, rather than enact a specific, official cultural image. And they note that different Candomblé groups challenge each others' practices, with accusations of producing only 'tourist Candomblé', bereft of legitimacy and power.

What they point to, then, on the one hand, is the relationship between the stress on specificity that is part of cultural tourism, and on the other hand, conflict about what those cultural groups are, what their attributes are and ought to be, and who has the power to define them and the authority to speak for them. These conflicts and processes of group identification seem to amount to a 'group-building' analogous to the nation building that, Holmes said, Belize abandoned. And if the case of Salvador is any indication, when the government of Belize sought to address the tensions and grievances associated with its melting-pot approach by changing to a policy of multiculturalism, it did not so much resolve them as shift them downward one level, from the nation as a whole to the groups that make it up.

Chapters 6 to 8 consider some of the aspects and consequences of cultural tourism, increasingly important in poorer countries. While government encouragement of this form of tourism may reflect social and political forces at work within a country, it is also the case that many see cultural tourism as a way to escape the competitive pressure of the commodity tourism of sun, sand and sea. While that commodity tourism may not stress particularity and cultural difference, it can still lead sets of people to put forward particular images of social groups, images that reflect and shape social and political interests and resources. Such images are the concern of Chapter 9, 'Tourism and its Others: Tourists, Traders and Fishers in Jamaica', by Gunilla Sommer and me, James G. Carrier.

Jamaica relies on commodity tourism, and the town of Negril, the focus of this chapter, is an example of what it can produce: a large number of all-inclusive, beach-front hotels owned by large corporations catering to foreign tourists. Not surprisingly, this has led to complaints in the country, as in the region generally, about the harmful environmental and exclusionary social effects of tourism. People in the tourism sector see in these complaints a challenge to their power to run their businesses as they like and to influence government economic policy. So, and again not surprisingly, they seek to defend themselves. This chapter looks at one aspect of that defence, the way that those in tourism describe themselves and others.

This defence revolves around the meanings assigned to different sets of people and to the activities associated with them. Those are the three groups in the chapter's title: tourists, traders and fishers. People in the sector defend themselves against the charge of social exclusion by portraying themselves as necessarily catering to the desires of tourists. They portray those tourists as pleasure-seekers who want to be pampered in a carefree tropical paradise, which means that they cannot encourage tourists to go outside the hotels' perimeter fences, for there they will meet undesirable Jamaican street traders who will harass them. They defend themselves against the charge of environmental harm in a different way, by saying that what local fishers do is even more harmful, and is the main cause of the degradation of the coastal waters.

These portrayals of tourists, traders and fishers have some truth to them. But they are perhaps more a function of the perspective and concerns of those who produce these portrayals than they are a function of the sets of people they portray. To begin with, they reflect and reinforce a set of cultural meanings that have long been important in Jamaica. In those meanings, poor Jamaican men, such as street traders and fishers, are seen as disreputable and heedless, while middle-class Jamaicans, such as managers in tourism companies, are respectful and conscientious. Thus, the meanings that they assign to themselves and others are particular manifestations of a set of values that portray and assign worth to different sets of people in this divided country, and so increase the power of some people rather than others to shape public life. As well, these renderings ignore the ways that the practices of the tourism sector itself shape the Jamaican world in which tourists, traders and fishers exist, and so shape what these people expect, want and do. In that way, they resemble the newer renderings of Blackness in Salvador: they construct sets of people whose attributes are seen as innate rather than being a result of the social settings and interactions in which they live their lives. And to the degree that these constructions are persuasive, they protect the tourism sector from responsibility for the state of affairs that excites the critics.

These chapters illustrate the different ways that tourism is associated with image and meaning. More importantly, however, they show how image and meaning are not just ideas in people's heads. Rather, they spring from social, political and economic relations, and because they shape how people see the world and each other, they affect social, political and economic interests. This means that if we are to understand the effects of tourism and its images, we need to place that tourism in the setting in which it exists. It is not just the growth of cultural tourism that affects the meanings and lives of people in Parara Paru, Belize and Salvador, their power to shape their own lives and the lives of others. Rather, tourism's effects are shaped by the tensions and concerns and differences in power that already existed: the distinct position of *Indios* in

Panama, the diverging interests of different groups in Belize, the long historical concern with Blackness in Brazil and the recurring class and colour tensions in Jamaica. To understand how tourism affects these countries, then, it is necessary to attend both to the changing nature and intensity of tourism and to the social, political and economic field in which that tourism exists, for these are related. Different groups in each country differ in their power to shape government policies affecting tourism and development; the changing nature and extent of tourism affects the power of different groups to shape government policy and public life.

Chapter 6
Tourists and Indigenous Culture as Resources: Lessons from Embera Cultural Tourism in Panama

DIMITRIOS THEODOSSOPOULOS

In the last 15 years, a small number of Embera communities located close to Panama City have succeeded in developing cultural tourism. These particular communities receive a regular flow of tourists and present them with certain features of Embera culture, such as traditional artefacts, music and dance performances. This recent development has encouraged those Embera who are involved to put their culture at the very centre of their self-presentation and their everyday activities. Following in the footsteps of their neighbours, the Kuna (see Swain, 1989), the Embera have realised that, in the context of tourism, their cultural identity as a people indigenous to the Americas, who still respect and practise their cultural traditions, can provide them with new and rewarding economic possibilities.

The particular encounters of the Embera with tourists have enhanced Embera cultural practices, and have demonstrated that Embera culture is desired and admired by Western visitors, who carry hard currency and are citizens of some of the world's most powerful nations. In this respect, tourism in Panama, like other cases examined in this volume (e.g. the chapters by Calvo-González and Duccini; Holmes), has played an important role in increasing the visibility of cultural diversity and in shaping the politics of representation. It has also inspired the Embera to re-evaluate their culture, and has provided new opportunities for them to enact and experiment with their indigenous identity and to relate to it in new ways. From their point of view, the new practice of entertaining tourists is an indispensable part of an 'authentic' and constantly evolving Embera culture (cf. Bruner, 2005).

Such a nuanced understanding of cultural authenticity, which is informed by a significant amount of anthropological work (e.g. Bruner, 2005; Coleman & Crang, 2002; Selwyn, 1996; Smith, 1989), can help us put in perspective the role of indigenous identities within the expanding economy of cultural tourism. The host-tourist interaction provides new, challenging opportunities for indigenous communities to exercise some degree of local control over their resources (Swain, 1989) and to rediscover, reflect upon and reconstitute their indigenous traditions (Abram *et al.*, 1997; Boissevain, 1996). More widely, it presents a contact zone for cultural exchange (Clifford, 1997), a meeting point of cultural expectations that influence and shape each other. The meaning of Embera indigenous identity is closely dependent upon this interaction of expectations, as well as a number of related practical and political circumstances.

Adam Kuper (2003), in an inspiring but controversial paper, has criticised the use of 'indigenous' in anthropology, which he sees as a more politically correct equivalent of 'primitive', but one that carries the discriminatory and crypto-evolutionary implications of the older term. Kuper, in turn, has been criticised by anthropologists who are concerned with the predicament of particular indigenous groups in under-privileged positions, and who see indigeneity as a politically useful concept that can contribute in the promotion of indigenous rights (see, among others, Kenrick & Lewis, 2004). In Latin America particularly, it is the indigenous actors themselves who have accepted the popular definition of the term, in an attempt 'to further their cause for ethnic recognition and self-determination' (see Ramos, 1998: 6–7). In Panama, the term 'indigenous' specifically describes Amerindian groups, such as the Embera, the Wouanan, the Kuna, the Kgäbe and the Bugle, who identify with the term and widely rely on it for political representation. In the economy of tourism, the term has obtained additional value and significance.

For the Embera, indigenous culture, as this is made visible and available in tourism, is a valuable economic and symbolic resource that has the potential to transform Embera identity and shape the politics of self-representation. Within the confines of their immediate community, those Embera who work in tourism become authors of their cultural performances and celebrate their indigenous identity with artistry and respect. They choose to focus their tourism presentations on visual and material aspects of Embera culture, which they see as more easily understood by the visiting tourists. Both the tourists and the Embera appear content with this type of engagement: the tourists consume a digestible amount of indigenous culture and enhance the cultural dimension of their holiday, while the local hosts make a good living without having to leave their community or compromise their indigenous identity. On the contrary, some of them explain, tourism provides them with new opportunities to practice their cultural traditions.

While this type of exchange between host community and tourists has many positive effects, it can hide from our view a set of social relations that are symptomatic of the wider economy of cultural tourism (see Carrier & Macleod, 2005). For example, the Embera, regardless of their skill in accommodating the needs and desires of their guests, are not in a position to control the flow of tourists coming from outside or to direct that flow among different communities. Embera settlements in inaccessible locations remain deprived of the benefits of tourism and perceive the tourists themselves as yet another valuable resource that remains beyond their reach. As this suggests, in this sort of tourism it is not just indigenous culture that is a valuable resource. The flow of tourists is too.

In the following sections, I will explore Embera cultural tourism as this takes place in Parara Puru, one of the communities in the Chagres National Park that receive tourists on a frequent and systematic basis. I will situate the success of Parara Puru as a site for cultural tourism within the broader context of a more general Embera desire to develop tourism, which for several other Embera communities remains unfulfilled. It is in this respect that indigenous culture, a resource available to all Embera, has emerged as the medium for claiming visibility, numbers of tourists and the material benefits of engaging with tourism, but has also created some imbalances between different Embera communities. The residents of more inaccessible Embera communities have access to one necessary prerequisite for the development of cultural tourism, indigenous culture, but not to the other, the tourists themselves.

Cultural Tourism in Chagres

The Embera are a people with a history of migration. When individual communities grow too large or when other problems occur, such as internal disagreements or external threats, they split and migrate through a network of interconnected rivers to new riverside locations, mostly in fairly remote and inaccessible rainforest environments. This is how they have spread from lowland Colombia, their original homeland in Choco, to the Darien Province just over the border in Panama, and further north in the isthmus to the rivers that sustain the Canal. It is this strategy of dispersion and migration that helped the ancestors of the Embera escape assimilation by the surrounding Latin American societies during colonial times (Williams, 2005) and has secured the survival of Embera culture since then. In the last 50 years, insecurity in Colombia has resulted in migration, not only across the border from Colombia to Panama (Kane, 2004), but also from Darien to locations much closer to Panama City.

It is those communities nearer the capital that have recently taken advantage of a new and unanticipated opportunity, the development of cultural tourism. This prospect has been systematically explored by Embera Drua, Parara Puru and Tusipono, three communities in the Chagres National Park, which are built, like most Embera communities, close to a river, in this case Rio Chagres. The inhabitants of these communities use dugout canoes to reach the closest latino towns on Lake Alajuela, which are connected with an asphalt road to Panama City. With those canoes, the Embera transport all necessary goods that they buy in these towns, and also the tourists who want to visit their communities. The latter are brought from Panama City to the edge of the national park in small buses. The proximity of these Embera communities to the capital undoubtedly plays a role in the success of their cultural tourism.

After a one-hour journey, the tourists are able to enter what looks from the outside like a remarkably different world: a landscape of rivers, lakes and rainforest vegetation, with small villages inhabited by Amerindians who emerge out of their thatched-roof houses dressed in traditional garb. This looks like a world lost in a primordial order of existence, a journey to a land that time forgot. What the tourists are not able to perceive is that the overall landscape has been affected by the construction of the Panama Canal and especially the establishment of the national park, which constrained Embera hunting and cultivation, their traditional subsistence activities. Instead, the Embera who live in the park were encouraged to develop cultural tourism and present the low environmental impact of their traditional culture. Communicating an eco-friendly message to visiting tourists would, it was thought, attract them and enhance their appreciation of the Embera landscape.

Images of the natural, both the rainforest and the indigenous people living in it, are highlighted in tourist advertisements. In guidebooks, tourist brochures and the internet, Chagres Park is advertised as an ideal setting for adventures in 'nature', including an opportunity to experience an indigenous Amerindian culture in a 'pristine', 'natural', 'tropical' environment. The typical tourist encounter is a two- or three-hour visit to the indigenous communities, which includes traditional music, dance, food and Embera artefacts (*artezania*) (see Figure 6.1). This standard

Figure 6.1 Embera dance performance

package can be supplemented by a walk in the rainforest, a swim in the local waterfalls or some informal instruction in Embera indigenous knowledge, for example of plants with medicinal properties.[1]

In all cases, Embera culture is at the very centre of the tourist encounter. The Embera hosts meet the tourists at the embarkation points, steer a course through the lake and the river, welcome the tourists to their community, offer an informal speech about the history of the community and the methods used to construct traditional artefacts, perform Embera music and dance, and paint a few tourists with traditional Embera designs using plant juice. They also encourage their visitors to stroll among tables covered with Embera handicrafts: baskets, masks and woodcarvings for sale at prices lower than in Panama City. The repetitive nature of these daily encounters with tourists has consolidated the expertise and professionalism of the local hosts (Theodossopoulos, 2007), who know how to handle both the representation of their culture and the visiting tourists with remarkable ease and self-confidence.

A Community of Tourism Professionals

Parara Puru, the particular community I study in Chagres, was founded seven years ago by Embera families already established in the area. The process of establishing their community does not differ significantly from that of other Embera villages in Panama Province or in Darien. Approximately 40 years ago, the Embera, encouraged by the government, gradually abandoned their traditional dispersed pattern of settlement to form concentrated communities (see Herlihy, 2003; Kane, 2004). They were rewarded with primary schools (one for each community), some provision of medical care and, more importantly, with the establishment of their own semi-autonomous reservation in Darien, the *Comarca* Embera-Wounaan, divided into the Cemaco and Sambu districts. Within these territories, the Embera were allowed a degree of self-government and political representation. The Embera living on lands outside the *Comarcas* also formed concentrated communities in Darien, and in various locations in Panama Province, such as the Chagres National Park. Parara Puru is one of the communities that lie outside the designated Embera reservations.

The founders of Parara Puru were predominantly born in Chagres of parents who had migrated from Darien 40 or 50 years ago. They were joined by others, who had moved from Darien more recently and who were related to the founders by family ties. They had chosen the location of the community after carefully evaluating the requirements of receiving and entertaining regular, small and large groups of visitors. For example, while the community was typical in that it was built next to the river, it was on a site where the river's flow was adequate for transporting tourists by canoe during both dry and rainy seasons. Villagers took special care to construct their dwellings as fine examples of Embera architecture, and they created special spaces for dance and artefact presentations. In most respects, their community looks very much like any other Embera community, but the landscape, the architectural style and the arrangement of the thatched-roof dwellings has been planned with special care and attention to detail.

In fact, the appearance and spatial organisation of the community is designed to pass a clear message to the visitor: Parara Puru is first and foremost an Embera community, where Embera culture is made available to outsiders to be consumed visually (Urry, 1995), but where it is also explained and respected. Within the community, as I will show, the Embera hold firm control of the tourist exchange, take care of their visitors and promote their cultural traditions. They expect a standard fee from every visitor for the food, hospitality and cultural presentations they offer, and share the profits fairly among the members of the community. They receive tourists daily through the year, and are prepared to entertain and present their culture to large groups of seventy or a hundred, or to small groups of three or four.[2]

The residents of Parara Puru admit that working with tourists is a demanding job, one that keeps them busy throughout the year and offers very few opportunities for vacations. However, they also recognise that the financial rewards of this new occupation are much higher than the conventional Embera activities of fishing, hunting and small-scale cultivation. Some residents also say that playing music, dancing and talking to tourists about one's own culture is more enjoyable than the hard labour of cultivation that they experienced before tourism. These people feel lucky when they compare themselves to Embera who live in villages in inaccessible areas not suitable for tourism. Work in tourism, they explain, allows them to continue practising their traditions and remain closely connected with their Embera identity, but without having to migrate to the city or confine themselves to poverty.

Indigenous Culture as a Valuable Resource

Tourism has brought about the re-evaluation of Embera culture. Until 15 or 20 years ago, the Embera were an indigenous people on the periphery of the Panamanian state, occupying lands unsuitable for intensive cultivation or systematic colonisation. Along with other Amerindian groups, they were stereotyped as *indios* (Indians) and occupied the bottom of the colonial, and later national, social ladder. Nowadays, the Embera figure prominently in advertisements for most national and private tourism initiatives. They appear in tourist pamphlets and prospectuses available to foreign visitors in airports, tourist offices, the market place or the internet. In these images, the Embera appear as representatively Embera as possible: women, men and children wearing traditional Embera attire, posing in front of traditional Embera architecture, handling Embera artefacts or engaging in traditional Embera activities. Like their famous neighbours, the Kuna, their culture has become emblematic of 'indigenous' Panama.

The presentation of the Embera in tourism has facilitated a shift in how they are understood, one that has progressively moved them away from the category of the *indio* and closer to *indigenas* (indigenous), a term associated with a certain degree of acceptance and an acknowledgement of rights. Nationally, the Embera are gradually becoming more widely accepted by the broader Panamanian society as an example of Panamanian cultural diversity or Panama's indigenous heritage. At the same time, this visibility in tourism, and the associated regular contact with foreign visitors, has made the Embera visible internationally: more people outside Panama are now familiar with Embera culture (most of them from the wealthy and powerful nations of the North) and their acquaintance with Embera culture carries an aura of recognition, admiration and legitimation.

This change in Embera status is recent and directly linked with the economy and politics of tourism. A generation ago, the Embera were referred to by non-indigenous Panamanians as *Chocoes* (the Indians from Choco in lowland Colombia). Their position in Panama was marginal, with most communities living in Darien, a province left deliberately under-developed to form a boundary of impenetrable, thick rainforest between Panama and Colombia, intended as a natural barrier stopping livestock epidemics and unwelcome migrants.[3] From Darien, the Embera spread further northwards in Panama Province, while some families reached the most north-westerly point of the present Embera geographical distribution, the river systems that support the water flow in the Panama Canal. This particular type of migration is, as I have noted, in accordance with Embera tradition. 'This is what the Embera do', my respondents underline.

When, late in the 1990s, the Embera began to get involved in tourism, they realised that they possessed a newly valuable resource, their culture. Traditional Embera architecture, attire, music and artefacts provided the setting, the inspiration and the consumables for the tourist exchange. What they already possessed – their knowledge about the world, their culturally specific adaptations, their very own traditions – was now in demand, and the demand was rewarding. In this way, indigeneity, the link with a distinctive culture native to the Americas, provided the Embera involved in tourism with new, unanticipated opportunities. While the changing approaches to indigeneity in the last 40 years meant that the wider Panamanian society was moving away from the dismissive colonial attitude and was prepared to recognise Embera culture and treat it with some basic respect, the more recent engagement with tourism introduced the Embera to a new and previously unparalleled level of recognition. Tourists from the wealthy nations of the North were now approaching their culture with admiration. They wanted to learn about and experience Embera culture, thereby elevating indigenous cultures relative to their more mainstream neighbours in the wider latino society. Cultural difference, embodied in the Embera, was an indisputable asset in tourism. Embera culture had value.

Tourists as a Resource

I have shown how the Embera of Parara Puru can earn a living from tourism simply by being themselves, by practising their distinctive cultural identity and traditions. Their indigenous identity, valorised by foreign visitors, is in demand, and the Embera who host tourists in their communities are learning, very fast, how to satisfy this demand. But in Panamanian cultural tourism, indigenous culture is not the only essential resource. From the point of view of the Embera involved in tourism, tourists are too. An example will illustrate this point.

One morning in February, at the peak of the tourism season, I was waiting with some of the residents of Parara Puru for the tourists to arrive. The canoes from Embera Drua, a neighbouring community up the river, were ascending the stream with difficulty on the low waters, heavy with tourists wearing orange life-jackets. We were staring at the canoes of our neighbours, all filled with generous numbers of tourists, confident that the canoes of our community were on their way, bringing a comparable load of visitors. This was a good time for Embera tourism in Chagres and expectations were high. The talk, while we were waiting, focussed on a comparison of Embera life in Darien and in Chagres, and more specifically on the relative merits of agriculture and tourism. 'The canoes in the rivers of Darien are heavily filled with plantains', I remarked. 'But... the canoes in Rio Chagres are heavily filled with tourists', observed my Embera interlocutor (see Figures 6.2 and 6.3).

Figure 6.2 The canoes in the rivers of Darien are heavily filled with plantains

Figure 6.3 But the canoes in Rio Chagres are heavily filled with tourists

The comparison was amusing to the line of men resting on that hot morning. Some of them were born in Darien, others in Chagres, but they all identified with their present community and its recent success with tourism. They said that Embera always live by the rivers, their dugout canoes always carry the productive resources of the community. So, while the sources of wealth might change, the processes remain the same. They continue to live by the rivers, adapting to opportunities, forming new communities, expanding when the prospects are favourable. In Darien, Embera dressed in t-shirts transport their goods along the river. In Chagres, Embera dressed in traditional attire transport groups of tourist. It is difficult to tell in which setting, in which economy, Embera cultural tradition reigns more supreme.

Yet, it is easy to distinguish between communities that have good access to tourists and those that do not. The communities in Darien, and other communities a long way from the capital, cannot attract significant and sustained numbers of visitors. Some remote villages in Darien do not receive tourists at all. And yet, many Embera in these less privileged locations desire to enter the economy of tourism. They are aware of the success of their fellows in Chagres, and they wish to reproduce the cultural presentation of the latter (Theodossopoulos, 2007). They do not have, however, a clear understanding of the procedures that rule the flow of tourists to particular sites, much less any sense of control over them. In other words, the Embera wishing to develop tourism in those relatively inaccessible locations possess the first prerequisite for gaining access to the resource that is tourism, for they possess Embera culture. However, they lack the second prerequisite, tourists themselves.

This unequal distribution in the flow of tourists is gradually differentiating Embera villages. Embera in the inaccessible communities of Darien are producing increasing quantities of traditional artefacts, which they sell to Embera in the tourism communities at Chagres. The latter are too busy entertaining successive waves of tourists to meet the tourist demand, and despite their efforts to produce some baskets, masks or carvings themselves, they have to rely on the supply of artefacts from more distant villages. Women from these more distant villages will make the long journey from Darien to the communities in Chagres to sell some of their handicrafts to Embera women living there.

Tourism was introduced in the Embera communities in Chagres fairly recently: seven years ago in Parara Puru, and in some neighbouring locations three or five years earlier. But its impact on the Embera of Panama is already significant. In my field trips in Darien, I became aware of the strong desire of other communities to develop tourism. In the last three or four years, some have been attempting to do so, but with little success so far (see Theodossopoulos, 2007). That is because the desire of most tourists to have an encounter with an indigenous Amerindian culture is not strong enough to overcome the prospect of bad roads, poor infrastructure and the arduous task of travelling long distances in rainforests. At the moment, Parara Puru and its neighbouring communities in Chagres, located closer to the flow of tourists, receive a fair share of the profits derived from cultural tourism. Other, more distant Embera communities have not been so lucky.

Authenticity and Respect

For people who have engaged with tourism only recently, the Embera of Parara Puru have entered the task with surprising confidence and success. They handle the available supply of tourists with care, respect and self-assurance, and expect equal respect from their guests. In the context of cultural performances for tourists, and within the immediate territory of the local community, the Embera hosts have emerged as skilful tourism professionals, secure about their indigenous identity and well aware of its value. During the course of their exchanges with the tourists, they closely control which aspect of their culture they wish to make available for display, which part of their identity to highlight in their self-representation.

The main focal points of those cultural presentations are Embera music, dance and material culture. The latter includes the *artezania* available for sale, priced and placed on tables, but it also includes the buildings and objects in the surrounding environment, such as thatched houses and dugout canoes. The Embera hosts have much to say about these aspects of their material culture, offering detailed descriptions of the raw materials and techniques used during manufacture. Other aspects of Embera culture, such as religion and curative practices, are less visible during the standard presentations for tourists. However, those tourists who are interested in such topics, and are prepared to spend some additional time and money, can easily arrange a more thorough private introduction by an Embera guide, expert in indigenous medicinal knowledge and related subjects.

In Parara Puru, and in other communities that receive tourists regularly, the community leaders greet tourists with a short introductory talk covering the history of the community and the circumstances that encouraged the Embera to engage with tourism. They also describe the traditional materials or techniques used for the manufacture of particular types of *artezania* and the amount of effort invested in their construction, which indirectly justifies the prices being charged. As the Embera explain, a handwoven basket or a mask that requires 20 days to make is priced at US$20, while another similar artefact that is completed in 15 days costs only US$15.

Questions about the history of the particular community are often inspired by discussions that tourists had before visiting Chagres, for example with other tourists who had already visited the Embera. An issue that troubles a few tourists is the authenticity of the communities in question. Some have come across similar indigenous performances in other parts of Central America or the Caribbean, while some others appear slightly uncomfortable with the well-rehearsed nature of the Embera cultural performances and the number of artefacts available for purchase. Having second thoughts about the authenticity of the cultural experience on offer, some inquisitive tourists ask questions that attempt to uncover the 'unreal', not properly 'indigenous', dimensions of life in the local community. For example, on more than one occasion, I have heard tourists enquiring about the small number of elderly people in Parara Puru, or the fact that Embera children are educated in Spanish.

From what I have seen, the inhabitants of Parara Puru are well prepared to answer those questions. For example, they explain that their community was founded only seven years ago, but that most local residents were born or arrived in the Chagres area several years before. They also clarify that it is part of the Embera way of life to found new settlements and to move to new locations during the course of one's life. This is why, my respondents in Parara Puru point out, some of their older relatives do not live in the particular community, but instead live in neighbouring ones, a few minutes' boat ride away. In their answers, the inhabitants of Parara Puru describe the practical dimensions of life as an indigenous people in the contemporary world. It is true, they admit, that their engagement with tourism, and the regulations of the Chagres National Park, have curtailed their involvement in traditional Embera subsistence activities. This is why they now buy most of their food from the nearby latino towns. It is also true that their children go to school and are educated in Spanish; but they speak Embera with their parents, and after returning from school they actively participate in the Embera cultural performances that take place, day after day, in the community.

In fact, one can argue that the children in Parara Puru are exposed to Embera cultural practices that are more traditional than those experienced by children in Embera communities that do not regularly receive tourists. It is therefore probable, as I have argued elsewhere (Theodossopoulos, 2007), that they will grow up to become individuals with a very explicit and particularly well-defined identification with Embera culture and tradition. In a similar manner, through their recurrent reproduction of Embera cultural traditions, the adult inhabitants of Parara Puru are becoming experts on all matters Embera. Through the confidence acquired by everyday enactments of indigenous culture, they engage with their culture in terms of a new, expert level of authentic involvement, which involves a certain degree of improvisation and the inscription of their own identities into an evolving Embera cultural tradition.

This type of personal and spontaneous involvement in staged performances allows us to question MacCannell's (1976) rigid distinction between the front- and back-stage of tourist exchanges. As Bruner (2005: 5) argues, the challenge for contemporary anthropological research on tourism is to move beyond 'limiting binaries' such as the 'authentic-inauthentic, true-false, real-show, back-front' distinctions. The performances for tourists in Parara Puru are examples of a new culture that develops around the theme of indigenous tourism. In this context, the Embera traditions presented to tourists are not simply rediscovered, but are put into everyday use and become an indispensable part of the lives of a people who structure their everyday routines around traditional practices. The culture performed for tourists, as Tilley (1997) has argued with reference to indigenous tourism in Vanuatu, is not merely reducible to marketing and commodification. In many cases, it represents an evolving tradition that is gradually adapting to new challenges and new circumstances.

The Embera of Parara Puru are tourist entrepreneurs who rely on their indigenous identity and a steady supply of tourists to make a living. They are expected to conform to a static perception of an unchanging indigeneity, while at the same time they try to adapt to the conflicting demands of the wider national society. Within the context of cultural tourism, they have to fulfil two parallel identities simultaneously, that of an indigenous people and that of citizens of a modern nation-state (cf. Robins, 2003: 398). The former corresponds to a representation of 'absolute alterity', a perception of 'a mythic, primordial Other' (Robins, 2001: 849). The latter necessitates a realistic, strategic adaptation to the particular circumstances of a competitive tourist economy, and the consolidation of their engagement with tourism within the wider context of the Panamanian economy.

Complex considerations like these can help us to understand why the standard narratives about Embera culture, as this is presented to tourists, do not include painful or explicitly political topics, such the politics of ethnicity or the experience of being a minority group. Embera cultural tourism, in its early stages of development, has been a meeting point of the Embera communities and members of the wider national community, providing opportunities for cooperation with the government, NGOs and tourist agencies. And at least for those communities that have succeeded in attracting good numbers of tourists, the result has been rewarding. In this respect, cultural tourism in Panama has provided a point of convergence, rather than of antagonism and conflict, between the interests of indigenous communities lucky enough to receive substantial numbers of tourists and the desire of the Panamanian government to encourage the development of new economic activities.

Conclusions

Through their recent engagements with tourism, the inhabitants of Embera communities in Panama that receive substantial numbers of tourists have emerged as dynamic protagonists in the tourism exchange, claiming their share of authorship in an economy based on the consumption of indigenous culture. Within the immediate vicinity of these particular communities, Embera hosts retain a strong sense of control over the presentation of their indigenous identity. They invite the tourists to focus on presentations of music and dance, and on material aspects of their culture, examples of which are available for purchase. The attention to these material and performative aspects of Embera culture appears to facilitate this particular type of indigenous tourism, as it directs the tourists towards concrete visual experiences or objects, whose value and cultural significance are easily communicated and safely consumed.

By focusing on particular cultural practices and artefacts, we might lose sight of the social relations that bring the tourism experience into existence (Carrier & Macleod, 2005: 330). The tourist 'bubble' excludes from the tourists' view 'the antecedents and the corollaries' of the tourists' visit (Carrier & Macleod, 2005: 316). What is interesting to note here is that the Embera of Parara Puru, like those in other tourism-oriented communities in Chagres, appear to benefit from the protective bubble of the standardisation of the tourist experience. They keep their interaction with the tourists uncontentious politically, while at the same time they improve their economic circumstances and gain valuable experience of the practical rules of the tourism industry. In the local context, the relative standardisation of cultural tourism packages that undeniably diverts attention from the broader social context of the production of Embera culture, appears to provide the host communities with a sense of control over the tourist exchange and a convenient starting point to structure their interactions with tourists.

Furthermore, and at a more general level, it appears that tourism has increased the visibility of ethnic minorities in Panama (cf. Guerrón-Montero, 2006; Tice, 1995). This visibility might be sanitised politically, but from the point of view of the Embera and other indigenous groups, it is much preferable to the previous disenfranchisement and stereotyping. Since the introduction of tourism, Panama publicly acknowledges Embera indigenous culture. National tourism campaigns promote ethnic diversity, and consequently shape the politics of ethnicity (elsewhere in the volume, Duccini and Gonzalez describe similar processes in Brazil; Holmes does the same for Belize). In Panama, the identification with an indigenous culture is slowly emerging as a process of empowerment and self-respect. The Embera, previously a peripheral and disadvantaged ethnic group, have now rediscovered the advantages of an indigenous identity, one that they had struggled to maintain in the past.

It is in terms of this cultural distinctiveness that the Embera communities in Chagres have been transformed into popular tourist destinations. The success of the particular type of cultural tourism they promote, together with its protective 'bubble' (see Carrier & Macleod, 2005), hides a broader disparity among the different Embera communities. The great majority of Embera settlements lie outside the affluent zone of tourism and can benefit from it only indirectly, by producing artefacts sold to villagers in the communities where tourists go. This powerlessness in controlling the flow of tourism has contributed to the development of a perception of the visiting tourist as a limited resource, a commodity that can benefit certain communities, but not others. In this respect, and in terms of the growing desire of the Embera to develop tourism, the tourists are as much commodified as the indigenous culture they consume. They are distributed between different communities, they are transferred in canoes along the river like plantains or other goods and they are treated symbolically like the valuables that fuel the local economy.[4]

In the world outside Embera communities, however, the flow of tourists remains beyond their control. It is regulated by the laws of supply and demand, competition between tourist operators, the periodic visits of cruise ships, the weather and tourist professionals who master languages other than Spanish or Embera. These practical considerations, as well as that of the accessibility of different villages, have privileged the communities at Chagres over other Embera destinations. These observations can help us understand the apprehension felt by some in the more distant Embera communities when they hear about the success of cultural tourism in Chagres. Although they can make equal claims to an Embera indigenous identity, they are excluded from the associated benefits.

In a previous paper, concerned with the politics of tourism in a Garifuna community in Roatan, Honduras, Kirtsoglou and I highlighted the loss of control experienced by indigenous communities in their attempt to promote their culture in tourism (see Kirtsoglou & Theodossopoulos, 2004). The local complaint that non-local agents were 'taking' Garifuna culture 'away' was closely related to the fact that visiting tourists, conceived once more as a valuable resource, were entertained outside the immediate locality of the indigenous community, and that the profits of the tourist exchange were appropriated by non-indigenous tourist entrepreneurs. This example, as well as the disparities between those Embera with and without access to tourists, can help us appreciate the importance of attracting tourists within the physical territory of the host community. When tourists are entertained within the immediate confines of the indigenous society, the consumption of indigenous culture by Western audiences does not necessarily contribute to a sense of powerlessness and loss of control. On the contrary, the local community can be empowered and retain a sense of authorship over its cultural representation.

The Kuna, the neighbours of the Embera, are a well-known example of this. Through active participation in the tourism business, Swain (1989) explains, they shape indigenous tourism and use it to support their own cultural survival (cf. Tice, 1995). In this respect, 'the Kuna can serve as a model for other groups' (Swain, 1989: 103). As in the case of political representation (Kane, 2004: 10), several Embera communities are already following the Kuna model. In Chagres, those lucky communities that receive a constant flow of tourists have accepted tourism as an everyday part of their lives. They now specialise in cultural tourism and are shaping the future of Embera tourism as they experiment with different ways of representing their culture to tourist audiences. Thus it is that the mundane practice of tourism transforms the Embera who participate in it, turning them into more experienced tourism professionals.

Without disregarding the emerging disparities in the wider political economy of indigenous tourism, it is important to underline the desire expressed by particular Embera communities to engage with tourism and, through tourism, with the wider world. From their point of view, their new involvement with the economy of tourism is nothing more than another conventional Embera adaptation strategy. They now work for tourism, the Embera hosts explain, but they do so in a well-established Embera way: they remain authentic in terms of social organisation and process, while they experiment with, refine and develop their cultural traditions in new directions. The culture of indigenous tourism, and some of the practices it entails, might be new, but they arise 'from within the local cultural matrix' (Bruner, 2005: 5). From the point of view of my respondents in Parara Puru, life is as authentic as ever. Dugout canoes heavily filled with goods, whether plantains or tourists, are travelling the Embera rivers, as in the past; it seems that it is still part of the Embera way of life to live by the river, to move along and to adapt to new opportunities.

Acknowledgements

I would like to thank the British Academy for supporting my research with the Embera and the editors of this volume for their inspiring comments.

Notes

1. Once in a while, a very small number of tourists who desire to escape temporarily from group tourism choose to remain in Chagres overnight. They depart on the second day with a subsequent group of visiting tourists.
2. Parara Puru residents distinguish a high and low period of tourist activity. From December to March, a period that coincides with the Panamanian summer or dry season, the tourist numbers are higher and the groups of visitors larger. After April, the numbers of visitors decline and remain low until August and September, with October and November being the rainiest months and the least suitable for tourism.
3. In the last two centuries, the Embera were permitted to cross and inhabit that boundary zone and share the lands of Darien with the Kuna, whom they partially displaced in their northwards migration, and the afro-Darienitas, one of the older resident groups of this inaccessible territory (see Kane, 2004).
4. In Djenné, a tourist destination in Mali, local guides use the analogy of 'hunting' to refer to the aggressive competition for tourists (see Joy, this volume). This is another example of perceiving tourists as a limited resource.

References

Abram, S., Waldren, J. and Macleod, D. (eds) (1997) *Tourists and Tourism: Identifying with People and Places*. Oxford: Berg.

Boissevain, J. (ed.) (1996) *Coping with Tourists: European Reactions to Mass Tourism*. Oxford: Berghahn Books.

Bruner, E.M. (2005) *Culture on Tour: Ethnographies of Travel*. Chicago, IL: University of Chicago Press.

Carrier, J. and Macleod, D. (2005) Bursting the bubble: The socio-cultural context of ecotourism. *Journal of the Royal Anthropological Institute* (N.S.) 11, 315–334.

Clifford, J. (1997) *Routes: Travel and Translation in the Late Twentieth Century*. Cambridge, MA: Harvard University Press.

Coleman, S. and Crang, M. (eds) (2002) *Tourism: Between Place and Performance*. Oxford: Berghahn Books.

Guerrón-Montero, C. (2006) Tourism and Afro-Antillean identity in Panama. *Journal of Tourism and Cultural Change* 4 (2), 65–84.

Herlihy, P.H. (2003) Participatory research: Mapping of indigenous lands in Darien, Panama. *Human Organisation* 62, 315–331.

Kane, S.C. (2004) *The Phantom Gringo Boat: Shamanic Discourse and Development in Panama*. Christchurch, New Zealand: Cybereditions Corp. (originally: 1994, Washington, DC: Smithsonian Institution).

Kenrick, J. and Lewis, J. (2004) 'Indigenous peoples' rights and the politics of the term 'indigenous'. *Anthropology Today* 20 (2), 4–9.

Kirtsoglou, E. and Theodossopoulos, D. (2004) 'They are taking our culture away': Tourism and culture commodification in the Garifuna community of Roatan. *Critique of Anthropology* 24, 135–157.

Kuper, A. (2003) The return of the native. *Current Anthropology* 44, 389–402.

MacCannell, D. (1976) *The Tourist: A New Theory of the Leisure Class*. Berkeley, CA: University of California Press.

Ramos, A.R. (1998) *Indigenism: Ethnic Politics in Brazil*. Madison, WI: University of Wisconsin Press.

Robins, S. (2001) NGOs, 'Bushmen' and double vision: The Khomani San claim and the cultural politics of 'community' and 'development' in the Kalahari. *Journal of Southern African Studies* 27, 833–853.

Robins, S. (2003) Comment on Kuper, the return of the native. *Current Anthropology* 44, 398–399.

Selwyn, T. (ed.) (1996) *The Tourist Image: Myths and Myth-Making in Tourism.* Chichester: Wiley and Sons.

Smith, V. (ed.) (1989) *Hosts and Guests: The Anthropology of Tourism.* Philadelphia, PA: University of Pennsylvania Press.

Swain, M.B. (1989) Gender roles in indigenous tourism: Kuna Mola, Kuna Yala and cultural survival. In V.L. Smith (ed.) *Hosts and Guests: The Anthropology of Tourism* (pp. 83–104). Philadelphia, PA: University of Pennsylvania Press.

Theodossopoulos, D. (2007) Encounters with authentic Embera culture in Panama. *Journeys* 8 (1), 43–65.

Tice, K.E. (1995) *Kuna Crafts, Gender, and the Global Economy.* Austin, TX: University of Texas Press.

Tilley, C. (1997) Performing culture in the global village. *Critique of Anthropology* 17, 67–89.

Urry, J. (1995) *Consuming Places.* London: Routledge.

Williams, C.A. (2005) *Between Resistance and Adaptation: Indigenous Peoples and the Colonisation of the Choco 1510–1753.* Liverpool: Liverpool University Press.

Chapter 7

On 'Black Culture' and 'Black Bodies': State Discourses, Tourism and Public Policies in Salvador da Bahia, Brazil

ELENA CALVO-GONZÁLEZ and LUCIANA DUCCINI

On 8 January 2007, the city of Salvador woke up to around 1500 portrait photographs of some of its citizens. These were scattered throughout the city in the 'Salvador Negroamor' exhibition, which presented itself as the 'biggest open-space photographic exhibition in the world'. Curated by the Brazilian photographer, Sérgio Guerra, the exhibition, according to its organisers, aimed to

> dialogue with [the city's] complex and contradictory reality, setting the basis for the necessary change in mentality so that the city can own up, without concessions, to its black and mixed identity, taking pride in it... using all the exotic African imagery, elected as fashionable by all layers of the city's population, to overcome discrimination against black and mixed persons. (www.salvadorne-groamor.org.br; accessed 7 March 2008)[1]

A visitor to Salvador arriving by air during the period of the exhibition would have been greeted, on the access road to the airport, with photos depicting *povo do Candomblé*, people whose costume associated them with the local Afro-Brazilian religion. When we commented on this to a member of the Department for Culture of the State of Bahia, we were told that these photos of *Candomblé* people 'truly represent the city's black population, which, in its majority, practices this religion'. However, the available statistics point to a very different reality. In Salvador, a city of approximately 2.7 million, 80% of people describe themselves as 'black' and 'brown', but only 0.49% declare their religion as *Candomblé* (FGV, 2005). Of course, there are some who declare themselves Catholic while at the same time being involved with *Candomblé* in varying degrees, and there are no reliable estimates for this group. Even so, there is still a considerable portion of the population who identify themselves as Protestants and who do not have current ties with *Candomblé*. What is at stake in the representation of the city as 'Black', and its Black population as linked to *Candomblé*, goes beyond numbers.

In this chapter, we want to consider how the city of Salvador, the first city to be founded in colonial Brazil, came to be known as the 'Black Rome', to be identified as the true repository of African culture within Brazil. Following Teles dos Santos' (2005) argument about the role of cultural politics in the re-signification of putatively African cultural practices as being the core of Bahian culture, we suggest that ideas about Bahian uniqueness, the basis of much tourism promotion, are linked not only to an idea of African cultural essence, but also, and with varying degrees throughout history, to an idea of a 'Black' biological essence, a 'Black body'. We also argue that the association between 'Black culture' and 'Black population', particularly with regard to Afro-Brazilian religious practices, was reinforced in the cultural politics of the Bahian state and was ultimately taken up by Black social movements, paving the way for the re-introduction of 'Blackness' as a base for particularistic policies in other realms of public policy, such as health.

What we want to show is how public policies designed to reach the 'Black population' carry a deep-seated tension between a concept of culture and an idea of bodily race, relying on the use of symbols like hairstyle, clothing and religious practices to express the 'true identity' of Blacks, that is, their 'Black essence'. In short, we present a case study of how the presentation and manipulation of imagery linked to 'Blackness' in public policies in the field of tourism feeds off historical discourses on culture and race, as well as contributing to the re-signification of the categories that are part of these discourses, thus facilitating their use in other realms of policy making.

Race, Culture and the Making of a Nation

The preoccupation in Brazil with what was considered to be its racial and cultural make-up can be traced to colonial times, when the miscegenation of the country's population was seen as a problem for its prosperity. From the perspective of biological racism, exemplified by the ideas of French theorist Gobineau (see Schwarcz, 1993: 13) and predominant in Brazil up until the beginning of the 20th century, miscegenation was condemned as the cause of the 'degeneration' of the country, pushing Brazil away from what many saw as its natural position as a civilised European nation. This reflected the fact that even though the ideas about human differences that emerged during the colonial periods of various European countries had diverse views of the significance of race, they nevertheless converged in the idea of European superiority (Wade, 1997: 9). Influenced by these French Enlightenment values, in which ideas about the nation are tied to the construction of an identity geared towards the future, Brazilian socio-political elites established an ideological differentiation between the civilised, aristo-cratic and superior part of the population, identified as whites, and the inferior, backward and non-civilised population, identified as the remaining segments of its population: indigenous peoples, Blacks and *mestiços*, those of mixed race (Skidmore, 1999: 67–73).

Debates often centred on the figure of the *mestiço*, who was seen in a number of different ways. For some, *mestiços* were deviants, incapable of civilisation and presenting a threat to the integrity of the 'white race' (as in the writings of the late 19th-century writers from the Bahian medical school, exemplified by Nina Rodrigues, 1899, 1988). For others, they were an undesirable but nonetheless necessary step in the path towards a whiter, more civilised nation (exemplified by the positions of Euclides da Cunha (1973) and Silvio Romero (1895), both also writing in the late 19th century). Yet for others, they were victims of an insalubrious environ-ment that did not allow them to develop their full potential for achieving a civilised state (see, e.g. Roquette Pinto (1920), writing in the 1910s).

For those who believed in the inherent impossibility of achieving a degree of civilisation with a predominantly non-white population, the solution came through a policy of 'Europeanisation and whitening', centred on the promotion of immigration from Europe. The European immigrants who arrived in the country represented the evolution of Brazil into a more industrial economy, since they were deemed to be more easily adaptable to a system of wage labour than were the freed population, seen as dependent and indolent. Crucially, these immigrants also represented the desire for physical whitening, which throughout Brazilian history had been tied to the modernisation ideal.

However, the phenotypic reality of the Brazilian population was far from the white ideal. Writing in the 1930s, Gilberto Freyre (e.g. [1933] 2003) acknowledged this fact and saw it as a reality to be celebrated. He pointed to a 'softer' slavery system, in comparison to that of the USA, and to the presumed harmonious coexistence of different races in Brazil, the so-called 'racial democracy', extolling racial miscegenation and the *mulatto* as the authentic characteristics of Brazil and of Brazilians. Freyre's work marked the spread of the idea that mixture lay at the core of Brazilianness, manifest in what Sérgio Buarque de Hollanda (1936) saw as Brazilians' cordial and flexible character. This does not mean, however, that the modernising project was dead. Rather, in a process that resembles what Holmes (this volume) describes for Belize, the state was to be responsible for moulding a national identity and an ideal future in which all Brazilians, separated by their different pasts, would have a place (Costa, 2002: 42).

This change in rhetoric did not, however, necessarily mean a radical change in common understanding, and some contemporary authors have pointed to the underlying acceptance of the whitening ideal in Freyre's project. For instance, Guimarães argues that 'racial democracy' came to signify the

> capacity of the Brazilian Nation (defined as an extension of European civilization in which a new race emerged) to absorb and incorporate *mestiços* and blacks. Such capacity required, in a tacit and implicit way, the agreement on the part of coloured population to snub their African or indigenous ancestry. 'Whitening' and 'Racial democracy' are, therefore, concepts of a new racialist discourse. (Guimarães, 1999: 53)

Similarly, other authors have argued that the shift in emphasis from 'race' to 'culture' was more a rhetorical than a conceptual change. Arthur Ramos, one of the representatives of the culturalist turn of the 1930s, argued that the problem with Nina Rodrigues' racialist analyses of the degeneracy of the *mestiço* in the 1930s was its use of the terms 'race' and 'miscegenation', a problem that, Ramos considered, could be trans-cended by replacing these terms with 'culture' and 'acculturation' (Ramos, 1939, as cited in Martínez-Echezábal, 1996: 111).[2]

However, the ideology of miscegenation was so successful in improving the public image of Brazil that a UNESCO project was set up in the 1950s, under the shadow of the Nazi genocide, to study Brazil in the hope that it could provide useful lessons for countries marked by racial conflict. A lot of this image had to do with Brazil's multiple and ambiguous racial classification system, which is said to be arranged around outward appearance, phenotypic differences and social position; that is, around 'marks' rather than 'origin' (Nogueira, 1955). This, for example, allows members of a single family to be classified in different racial categories based on their different appearances. The ideology of miscegenation does not, however, eliminate only the idea of a common ancestral matrix as a way to define racial groups and place people in them. In addition, it leaves no room for the idea of a culturally distinct

'Black community' composed of those who identify themselves as 'black' and 'brown'.

Even when some of the studies arising out of the UNESCO project contradicted the idea of Brazil as a racial paradise, the ideology of miscegenation remained strong. Thus, for example, the military government stopped including a race/colour question in the 1970 national census. The government argued that Brazil had no racial disparities and conflicts, and that distinguishing people on the basis of 'race' was actually illegal (racism became a felony in the 1950s), even though some researchers were going back to using 'race' as an analytical tool to identify inequalities.

The late 1990s saw a resurgence of 'race' in academic writings analysing social inequality in Brazil, as well as in public policies, particularly with the implementation of affirmative action policies in several state and federal universities. Central in debates about these policies are the questions of the existence and the nature of a 'Black community', as well as the impact that such policies can have on a country that imagines itself as mixed, both biologically and culturally. In these debates, both camps make arguments regarding bodies: the need to discern who can and cannot be considered 'Black', as well as whether 'browns' should be included as part of the 'Black population'. They also make arguments about culture: the existence of distinctive attributes that could help define the 'Black community' supposed to benefit from these policies.

Regardless of the merits of the particular arguments in this debate, their overall effect is to define the intended beneficiaries of affirmative action in a mixture of cultural and racial terms. The debate thus ends up reproducing the blurring of ideas of race and culture that feature throughout Brazilian history, while at the same time feeding, as well as reflecting, contemporary debates on multiculturalism and the place that it has in the building of the nation.

Blackness for Tourism

As we pointed out, at the end of the 19th century, the Brazilian 'Black' population became a central issue in politics, not as actors in their own right, but as a problem that should be dealt with. If 'Blacks' were initially the object of the medical gaze, by the 1930s it was their culture that was at stake, with 'Black' religion being seen as its unique and defining feature. Some of the first Brazilian social scientists, as well as many foreign anthropologists, turned to the religion called *Candomblé* as central to the expression and reproduction of that culture.

The work of Bastide (e.g. 1983, 2001) illustrates the way that studies of *Candomblé* were not linked just to the racial-cultural category of Blacks, but also to modernity (or its absence) and to some parts of Brazil more than others. This is apparent through his comparison of *Candomblé* and *Macumba*. He noted that Bahia had a bigger 'Black' population than São Paulo and Rio de Janeiro, and that Bahian *Candomblé* had more 'Black' followers than São Paulo's and Rio's *Macumba*.[3] Hence, he concluded, *Candomblé* was much more African, which is to say traditional, than *Macumba*. Bastide's work resonated with and fed into a decreasing celebration of tradition and an increasing concern with change and modernisation. 'Blacks' were seen as traditional, with an ancestral orientation, which meant that they were less adaptable than other groups in the country. What, then, was Brazil to do? How could it develop and become modern when it had such a large number of people seen to be so traditional? This problem was especially acute for Bahia, which was seen as almost totally 'Black'.

In his book about cultural policies in Bahia, Teles dos Santos (2005) shows how, ever since Vargas' dictatorship in the late 1930s, the model of de-centralised industrial development has been linked to the idea of profiting from local particularities for economic growth (this view still guides public policies for tourism in Bahia: see Governo do Estado, 2005). Since Bahia was seen as 'the Black state' of Brazil, the idea was to turn 'Black' culture, folklore and everyday life into the basis for the development of a tourism industry.

Our focus on national policies and debates might give the impression that *Candomblé* adherents were passive. This is not the case. They were involved in the disputes we have described, and invoked various notions current in scholarly and political debates to claim legitimacy and gain visibility, respect, allies and resources (Dantas, 1988). This echoes the case of the Embera (Theodossopoulos, this volume), where indigenous people reframe their identities in the context of tourism. This sort of reframing can be profitable for the groups that undertake it, and can be linked to issues like social rights and citizenship, as Holmes (this volume) describes.

Tourism was not the only economic area where Black culture and identity were to become important. Early in the 1960s, under president Jânio Quadros, cultural and economic policies became closely associated:

the affinity with the populations of the African continent, whose "culture marked our formation" resulted in a political solidarity, having as a tacit agreement the overcoming of the disagreements between the countries that formed what was to be known as the "Third World". (Teles dos Santos, 2005: 38)

The recognition of Brazil's deep roots in Africa meant that it became possible to base economic policies on a 'cultural' ground, including policies of national development. As a reflection of this, during Quadro's administration, some 'Black' cultural elements were gradually drawn into official policies, lending them an aura of Brazilian authenticity (Teles dos Santos, 2005: 54). This was extended during the military dictatorship, and gradually more things identified as 'popular manifestations' of culture and folklore began to receive public resources. In Bahia, this meant support, albeit limited, for handicrafts workers, *mestres de saveiros e de capoeira* (captains of a certain type of boat and leaders of *capoeira* groups), African-Bahian cuisine and *Candomblé*.

This changing government stance did not, however, mark a straight-forward recognition of African roots in Brazil. Rather, it resembled what Holmes (this volume) describes of Belize government policy before 1990, when ethnic diversity was seen as a feature of the country's past, and unity a feature of its present and future. Thus, in the view of the Brazilian government, things like *Candomblé* deserved support because they were not simple expressions of an African past. Rather, through Brazil's 'racial democracy' and racial-cultural miscegenation, they had become dis-tinctly and truly Brazilian. It was only in the late 1970s that some authors in Brazil turned to stressing distinctly African elements in Brazil's cultural mix.

This ambivalent view of things African was especially important in Bahia. Since the beginning of the 20th century, its image and identity have been built around the idea of being the 'Black state' of Brazil. This meant, for earlier racist theories, that it would be almost impossible to develop the region and make it modern. But, as time passed and the culturalist approach became important, what had been a handicap became valuable. Being 'Black' could make the state attractive to domestic and international tourists. The government decided to make this Blackness more visible and interesting.

The general idea behind public policies of tourism and culture in Salvador city seems to reflect the intertwining of preservationism and 'Blackness'. In Brazil as a whole, ever since the military dictatorship, governments have supported the maintenance and restoration of cultural and historical sites, mainly in those cities that are seen as potential tourist attractions. This has led to various actions to keep historical sites as they 'originally' were, through the renovation of old buildings and the support of folklore groups and their performances. This focus on what is taken to be original has meant that different places have become associated with particular phases in their history. Hence, cities in Minas Gerais, such as Ouro Preto, Mariana and São João Del Rei, are seen as living portraits of the gold-mining time. Other cities, like Gramado in Rio Grande do Sul, must be kept as testimonies of German immigration, allowing them to be 'traditional' and 'modern' at the same time. São Paulo is the industrialised city, and therefore what must be preserved are the railways and the buildings of the beginning of the 20th century. Salvador is the city of colonial splendour, with buildings and sites from the 18th and 19th centuries as the most important places. Many of these are, of course, associated with slavery.

Pelourinho (see Figure 7.1), the old central neighbourhood in Salvador, presents an interesting case. This tourist site helps to shed light on how Salvador became the 'Roma Negra'. The city originated in *Pelourinho* and the harbour area just below it, where there is the *Mercado Modelo*, another 'picture postcard' from Salvador. The area is home to seven important Catholic churches, with two more close by, as well as the archaeological remains of the first Brazilian cathedral, *Catedral da Sé*. While the area's streets and architecture reflect the colonial past, the image of that past that is remembered and presented hardly mentions Portuguese settlers and Brazilian Indians. In Bahian history, both popular and official, what is remembered is the slave workforce deployed in the building of colonial cities.

Figure 7.1 A square in Pelourinho

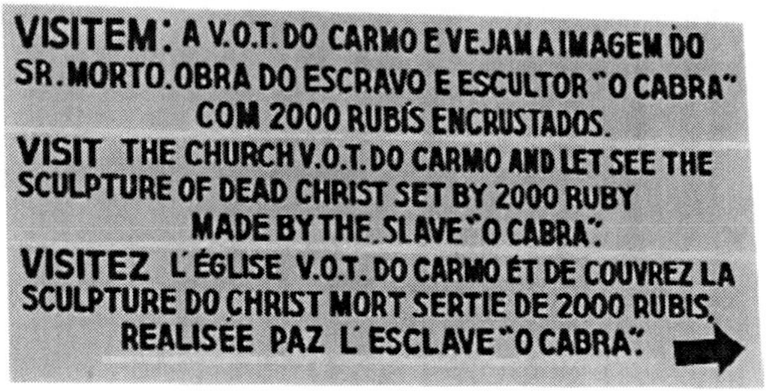

VISITEM: A V.O.T. DO CARMO E VEJAM A IMAGEM DO
SR. MORTO. OBRA DO ESCRAVO E ESCULTOR "O CABRA"
COM 2000 RUBÍS ENCRUSTADOS.
VISIT THE CHURCH V.O.T. DO CARMO AND LET SEE THE
SCULPTURE OF DEAD CHRIST SET BY 2000 RUBY
MADE BY THE SLAVE "O CABRA".
VISITEZ L'ÉGLISE V.O.T. DO CARMO ÉT DE COUVREZ LA
SCULPTURE DO CHRIST MORT SERTIE DE 2000 RUBIS,
REALISÉE PAZ L'ESCLAVE "O CABRA".

Figure 7.2 Sign describing o Cabra sculpture in the Museu do Carmo

For instance, in the *Museu do Carmo*[4] there is a sculpture of a crucified Christ, whose blood is said to be made out of 2000 rubies. The information sign (see Figure 7.2) that describes the piece makes no effort to place it in the artistic context of Catholic sculpture. Instead, visitors are informed that the statue was made by a 'slave and sculptor' known as *o Cabra* (a term used in the 18th and 19th centuries to refer to the offspring of a 'black' and a 'mulatto'). Something similar happens in the tourist cities of Minas Gerais, famous for their churches and for the sculptures by Aleijadinho. Whenever we hear about them, we are reminded that they were built by slaves and that the famous sculptor was the son of an architect and a slave. This stress on slaves is, moreover, localised in Brazil. In contrast to Salvador, the presentation of São Paulo's history stresses the indigenous and Portuguese people much more than it does the slaves (only 'Blacks' were slaves in Brazil). This is exemplified by the image of the Bandeirante, the first outsiders to conquer the area's hinterlands. They are described as *mestiço*, but arising from the union of 'Indians' and 'whites'.

Tourists in Salvador are constantly reminded of the city's 'Blacks'. The name of the main tourist neighbourhood, *Pelourinho*, is taken from the place where slaves were punished. 'Black' Salvador is presented in tourist postcards and commemorated in the names of streets. The 'Black' *orixá* cult appears in statues in places like the city's main post-office and in the names of apartment buildings, even in upper-class areas of the city. At *Mercado Modelo*, one can watch a *roda de capoeira* (*capoeira* performance), have a *moqueca* for lunch (a local fish stew with coconut milk and palm oil) and visit the basement where newly arrived slaves were kept before they were sold. In this visibility of 'Black' people in Salvador, we see how the country's stress on culture is linked to tourism as an economic strategy.

Although, as we have noted, the stress on 'Blacks' is limited to certain parts of the country, that stress is part of a general policy that sees tourism based on a cultural ground as a route to the economic development of poorer areas. Popular culture was worth preserving, to a significant degree, because it could attract tourists, resources and development.[5] This policy has two main consequences. Firstly, the manifestation of popular culture becomes a subject of government concerns and policy. Secondly, popular culture and tradition have become seen as things to be preserved exactly as they had 'always been'. Since tourism policies were linked to cultural ones, they have shared the same notions and ways of action, that is, supporting and reinforcing what is seen as 'traditional' or 'authentic', while disregarding the deep tension between 'race' and 'culture', purity and miscegenation, which has been active throughout the history of Brazilian social thought.

An interesting example of this concern with the preservation of unchanging cultural elements is the case of the *baianas de acarajé*, street sellers who produce and sell bean paste fried in palm oil, something deemed a local delicacy. This street selling goes back to slavery times, when the *baianas* used to sell *acarajés* not just to support themselves (or their owners), but also to fund *Candomblé* activities. In 1965, the recently created Superintendência de Turismo de Salvador (SUTURSA, the municipal governmental body in charge of promotion of tourism) dictated that only *baianas* wearing the full traditional attire, which is very similar to the clothes worn by active participants in *Candomblé* ceremonies, could be allowed to sell *acarajé* in the streets. The idea was to promote and rescue the use of the 'traditional' dress (see Figure 7.3), standardising the image of these street food sellers. However, the measure was not particularly welcomed by *baianas* themselves, some of whom complained about having to walk through the city to their stalls while wearing the required outfit. The solution proposed by SUTURSA was to provide rooms in the city centre where the *baianas* could change into the traditional clothes before setting up their stalls (Aquino de Queiroz, 2005: 345). The preoccupation with *baianas'* attire came up again late in the 1990s, this time in the shape of a municipal law that, among other things, required *acarajé* street sellers to wear the traditional dress (Reinhardt, 2006: 102).

As this concern with tradition suggests, *acarajé* is not simply a conventional food. In 2004, it was declared to be a 'cultural patrimony of Brazil', an official government designation. This led to controversy

Figure 7.3 *Baianas de acarajé*

over the sale of *acarajé* by evangelical *baianas*. The Instituto de Patrimonio Histórico Artístico Nacional (IPHAN, the government body in charge of preserving the country's historical and artistic patrimony) published a decision in favour of a request presented to it by the president of the Association of Baianas de Acarajé, the religious head of the Ilê Axé Opó Afonjá (a *Candomblé* temple known for its concern to reinstate what it saw as purer, more truly African *Candomblé* practices) and the Centre for Afro-Brazilian Studies of the Federal University of Bahia. In the decision, the IPHAN says that *acarajé*

> cannot be separated from its sacred elements, as well as from the elements that are associated with its sale, such as the complex attire of its sellers, the shape of the stall, as well as the location where the stalls are mounted... [which] remembers the old places where slaves used to sell food during colonial times. ... The traditional practice [of the commerce of *acarajé*] is under threat by the new phenomenon of the sale of *acarajé* in bars, restaurants and supermarkets, as well as its appropriation by other cultural universes, such as the version known as *"Acarajé* of Jesus", sold by members of evangelical religions. (IPHAN, 2004: 2–3)

The case of the *acarajé* points to a more general tension that is not, and perhaps cannot, be resolved. On the one hand, popular culture is valued because it is seen as authentic and unchanged. On the other hand, that culture is used for commercial purposes in tourism. This commercial use is likely to corrupt the very thing that preservationist public policies try to protect. This tension remains, creating both conflict and affinities between government officials and popular culture producers.

Salvador became the 'Black city' not only because the majority of its population is dark-skinned. After all, there are lots of 'Blacks' in Porto Alegre, in the state of Rio Grande do Sul, although obviously not on the same scale, but that city's identity remains *gaúcha* (a mix of Guarany Indian and Spanish, with Portuguese and Italian added). In addition, Salvador became a 'Black city' in part because being such mattered for the development policies designed for Bahia. As we have pointed out, these policies were based on the notion of local specificity, which meant that Bahia should profit from both its natural beauty and its unique culture, which is thought of as African-Bahian. The importance of this image for the state, and especially for Salvador, is apparent in the ubiquitous *Lonely Planet* guide:

> Much of Bahian life revolves around the Afro-Brazilian cult Candomble, and much of the color, symbolism and language of Bahian culture stems from its traditions. For example, the colored beads many Baianos (inhabitants of Bahia) wear about their necks represent their protectorate orixas (deities). The band name Olodum is Yoruba (an African dialect) for 'God of gods.' So to deepen your understanding of your surrounding, spend an evening at a terreiro [temple] witnessing the ritual and possession of a Candomble ceremony.
>
> Activities usually start around 8 pm or 9 pm any day of the week, in neighborhoods far outside the center. Bahiatursa or the Federacao Baiana do Culto Afro-Brasileiro can provide schedules and addresses for some terreiros, or Singtur can provide guides. If you go without a guide, make sure there will be taxis for the return trip. Wear clean, light-colored clothes (no shorts) and go well fed as ceremonies last for hours. (*Lonely Planet*, 2005: 423)

Being Afro-Bahian, however, presents a tricky problem. If it is not to contradict the image of racial democracy that is important in Brazil, it should be seen as *mestiço*, authentically Brazilian, rather than African. This problem has not been solved, and might never be, because it carries two reified notions that we pointed to at the beginning of this chapter, culture and race.

What is thought of as 'Black heritage' is important in Bahian identity and valuable in tourism. However, it is not clear exactly what the phrase means. This heritage does not bring us back to 'Africa', and neither does it refer to a specific group. 'Black' people in colonial Bahia came from many different places in Africa, and even the categories used at the time to classify them were fluid (see Parés, 2006). Slaves brought from Africa spoke different languages. They had different religious practices, some being Muslim and some worshipping local gods. They came from places with different social and technological systems. Nevertheless, in Brazilian thought, Africa became a huge, uniform place. In *Candomblé* studies, this idea of a uniform Africa is referred to as the 'mythical Africa' (see Birman, 1997; Capone, 2004; Dantas, 1988). As the notion of 'real cultural experience' is an important basis of cultural policies and is important for attracting tourists to Bahia, that experience needs to maintain its relation to 'Africa', to its 'deep roots'. But because there is no such thing as 'the' Africa, the legitimacy of Bahian claims to African roots became a matter of dispute, in which government personnel, scholars, Black movement members, artists and religious people take part. If even 'everyday Bahian life' is a tourist attraction because of its particular 'Black roots', it is important that everyone can acknowledge Bahia's distinctiveness. But then we have to ask, what is this 'Black' cultural distinctiveness?

For cultural policies in Bahia, it does not matter that many 'Black' people today belong to Pentecostal churches, study at universities, become professionals or belong to the middle classes. Such people are irrelevant because they do not convey a specifically 'Black image'. Rather, what is stressed in cultural and tourism policies is what is 'ethnic', regardless of whether or not those in the ethnic category have a sense of shared origin, community of experience or expectations for the future. This does not, however, mean that 'ethnic' is only a label to sell consumer goods, such as clothes, food or hairstyles or touristic experience. Rather, if we look closer, we can see that it refers to something that has always been present in Brazilian thought, the idea of a 'black race'. 'Black' people in Salvador in 2007 may have few shared experiences, but they have in common the phenotypic traits that identify them as such.

A Shift in Discourse: 'African' Cultural Practices, the Black Movement in Bahia and the Field of 'Black Health' in Salvador

As we have shown, the construction of Bahia's representation as 'the' 'Black' city owes much to cultural policies geared towards the promotion of tourism. These policies were effective at promoting Bahia's 'difference' as a tourist attraction: 4.1 million tourists visited the state of Bahia in

2000 (Aquino de Queiroz, 2005: 426). But aside from the economic, what sort of impact did this cultural policy have? In this section, we focus on the effects of this culturalist turn on public policies in Bahia, and chiefly the creation of a Black Population Health programme. We argue that the creation of such a programme is facilitated by the ideas that were reinforced by cultural and tourism policies, particularly that Salvador has a large 'Black' population. This construction of the city in terms of the idea of a 'Black culture' gets translated into a biological idea of a black race, expressed in the design and implementation of some aspects of the 'black health' policies.

The Black Population Health Group (BPHG) appeared in 2005, resulting from local and national efforts to implement health policies specific for the 'Black population'. Urged by activists from Black social movements, in particular feminists working on women's health issues, the BPHG reflects a political context sympathetic to combining universalism with some policies loosely based on a concept of multiculturalism that aims to address the needs and specificities of certain groups (see Chor Maio & Monteiro, 2005).

Thus far, the BPHG has mainly been involved in training health professionals in issues related to sickle-cell anaemia; gathering epidemiological data on disease and mortality with regard to 'race'; implementing the Institutional Programme Against Racism, a programme aimed at health workers that is funded in part by the British Department for International Development; and, lastly, the organisation of health promotion fairs in local *Candomblé terreiros*. What is especially interesting about these activities is the way in which the idea of a 'black race' as a biological entity is enmeshed in its rhetoric and practices with culturalist ideas of 'Black' identity.

Those urging the creation of the BPHG pointed to the high percentage of 'Blacks' in Salvador's population, as well as national statistics showing racial differences in morbidity and mortality. They also raised issues of discrimination, for example arguing that sickle-cell anaemia, a disease that is considered by both BPHG activists and some medical professionals as a 'Black disease', does not receive the same level of funding for research and treatment as other diseases, for the very reason that it affects mainly the 'Black Population'. A historical link is also made by these activists between current 'Black' health problems and the ill-treatment of slaves. The idea of a 'Black community' is often invoked, both in their appeals for solidarity ('our diseases were disregarded because they were affecting us') and by their references to a notion of shared cultural practices and 'ancestrality', embodied in the idea of

'Black religion'. Their rhetoric implies a biological basis for the 'Black population', expressed in things like their assertion that diseases such as diabetes, hypertension and sickle-cell anaemia are characteristic of the 'black race', phrasing that is used in some BPHG materials.

The most controversial BPHG activity has been that of health fairs in *Candomblé terreiros*. There was little controversy about what staff at these fairs actually do: measuring blood pressure, vaccinating animals for rabies, advising on dental health and nutrition (advice perhaps undermined by the presence of a *baiana de acarajé* selling her fatty, fried bean paste and sugary desserts). Rather, the controversy arose from the BPHG's association of 'Blackness' and *Candomblé*. The assumption, as we noted previously, is that *Candomblé* is the religion that best represents 'Blacks'. It links 'Black' people to their ancestrality, which is presented as beneficial: in the words of the coordinator of the BHPG, 'when people cultivate their ancestors, you see a positive impact on their health'. *Candomblé* is presented as a repository of 'Black' culture and ancestral knowledge on health (though the exact specificity of this knowledge is uncertain: see Chor Maio & Monteiro, 2005: 429).

Local evangelical politicians challenged this association between 'Blackness' and *Candomblé*, on the grounds of it being discriminatory. At a meeting of the municipal council, one councillor who agreed with special health policies for the 'Black' population (thereby pre-empting accusations of racism), disagreed with the organisation of health fairs in *terreiros*, arguing that 'in a city in which the population is mostly black, if one goes to a *terreiro* one is going to see a majority of blacks, but so is the case if you go to a Catholic church, or an evangelical temple'.[6] This association of Blackness with religions like *Candomblé* also occurs at the national level. The Technical Committee on Black Population Health, part of the Brazilian Ministry of Health, urges 'recognising the temples of religions of African origin [*religiões de matrizes africanas*] as spaces for the promotion of health', and mentions no such activities related to any other religion (Ministério da Saúde, 2004: 74).

It is interesting that, both in Salvador and nationally, the implementation of policies targeting 'Black Population Health' arose from pressure by organised Black movements. These movements, according to Teles dos Santos (2005: 164–165), were initially suspicious about the association of 'Blackness' and *Candomblé*. Early Black militants saw *Candomblé* as reflecting existing power relations, embodied in the influential 'white' members who linked important *terreiros* with local political structures. This began to change in the 1980s, when the movement began to adopt culturalist politics and accept *Candomblé* as an integral part of 'Black identity' to be incorporated into Black movement discourse.

BPHG policies and programmes, then, show the continuing overlap of the idea of 'Black' as a biological race and as a cultural group. There is certainly no direct link between the emergence of bodies like the BPHG on the one hand and the emergence of tourism-development policies on the other hand. However, they both reflect a continuing stress in Brazil on difference as racial, cultural or both.

Conclusion

From the 1950s onwards, tourism in Bahia rested on an idea of Bahian specificity, and the idea that the cultural manifestations that tourists experienced were authentic representations of everyday life. However, as we have shown, the everyday life presented to tourists was carefully crafted to include certain things and to exclude others, to promote the image of Bahia as a 'Black Rome', as a centre of 'Africanness' within the Brazilian nation. As such, it would be an exotic place, one worth visiting. Much tourism material makes reference to Bahia's 'magic', pointing to both the charming ways of its people, considered as 'happy' and 'welcoming to visitors', as well as the more muted quality of 'magic' within *Candomblé*. In the words of the *Lonely Planet* guide, you will only grasp Bahia once you visit a *Candomblé terreiro*.

The exotic magic of *Candomblé* in this Black Rome is the authentic Bahia that, tourists are told, needs to be experienced. However, as we have shown, the notion of authenticity can be slippery (Wang, 2007) and its performance can be shaped by much more than a simple desire for tourists and their money. To focus attention on only this desire and its consequences makes it hard to see more than what Grünewald (2003: 148) calls the 'touree', 'an actor who alters his behaviour seeking to profit from the perception that he is attractive for the tourist'. The situation we have described, alternatively, shows how notions of what is authentic can be appropriated and re-worked by people and state bodies with no direct involvement in tourism. The conflict around the *baianas de acarajé* illustrates the ways in which the idea of authenticity may become part of struggles for representation and legitimacy by those of different religious groups. The same is true of concerns about '*Candomblé* for tourists', which is not 'serious' and has no real *axé* (power). Here, the concern for authenticity, though cast in terms of tourists, is mostly about something else: the relative prestige and authority of different *Candomblé* groups and practices. The ideas of authenticity deployed by those in these different settings rest on an image of a shared culture, while at the same time shaping that culture to fit their differing interests and projects. 'Authenticity' thus becomes a marker and a weapon used in the political disputes.

At the same time, looking at tourism can help shed light on other areas of policy making. As we argued, the field of tourism promotion both built on and reinforced the idea of the existence of 'authentic' 'African' practices, or, in its translation into 'race', a reified notion of 'Black culture'. This reinforcement helped, in turn, to bring about bodies like the BPHG and their racially and culturally specific programmes and policies. Thus, it is that, at least in this part of Brazil, the concerns for authenticity that feed into tourism promotion can thereby be strengthened, and feed back in surprising ways into other, seemingly unrelated parts of social life and government policy.

Acknowledgements

Elena Calvo-González would like to acknowledge the support of the Conselho Nacional de Desenvolvimento Científico e Tecnológico of Brazil for the research that resulted in her contribution to this paper.

Notes

1. Unless otherwise noted, all translations are by the authors.
2. A similar shift in terminology, but continuity in concepts, appeared after the UNESCO Declaration on Race of 1950, in the field of biological anthropology. There, the category 'race' was replaced by that of 'population' (Ventura Santos, 1996: 132), and as we shall show, this category was used in some of the particularistic policies that aimed to target 'Blacks'.
3. Around the beginning of the 20th century, there was a substantial influx of Japanese and European workers into São Paulo and its 'Black' history was almost completely erased from popular memory. For instance, most inhabitants of São Paulo have never heard that Bexiga, which they know as the Italian neighbourhood, was originally a 'Black' settlement and that 'Blacks' there had practiced a kind of dance-fight very similar to *capoeira*, called *pernada* (Borges, 2001).
4. The Catholic Church rented the Carmo monastery to a Portuguese company, Pousadas de Portugal, which renovated the building extensively, transforming it into a very fancy (and expensive) hotel that combines antique architecture and art with luxurious, modern facilities and accommodation. Meanwhile, the associated church and museum at Carmo are almost falling apart.
5. It was only in 2007 that the tourism board of the Salvador city government was separated from the culture board. Those in the culture sector had complained that the former arrangement led to disproportionate amounts of money being put into tourism.

6. One of us asked a member of the BPHG whether they were organising activities in other religious settings, such as evangelical churches. The member replied that the conflict between *Candomblé* and evangelicals made it difficult to organise such events, thus supporting our hypothesis that, if the BPHG considers these conflicts as a difficulty for the implementation of actions with evangelical groups, it is because they see themselves as linked, at least symbolically, to *Candomblé*.

References

Aquino de Queiroz, L.M. (2005) A Gestão Pública e a Competitividade de Cidades Turísticas: A Experiência da Cidade do Salvador. Doctoral thesis, University of Barcelona.

Bastide, R. (1983) Cavalo dos Santos: Esboço de uma Sociologia do Transe. In R. Bastide (ed.) *Estudos Afro-brasileiros* (pp. 223–323). São Paulo: Perspectiva.

Bastide, R. (2001) *O Candomblé da Bahia.* São Paulo: Companhia das Letras.

Birman, P. (1997) O campo da nostalgia e a recusa da saudade: Temas e dilemas dos estudos afro-brasileiros. *Religião e Sociedade* 18 (2), 75–122.

Borges, R. (2001) *Axé, Madona Achiropita: Presença da Cultura Afro-Brasileira nas Celebrações da Igreja de Nossa Senhora Achiropita, em São Paulo.* São Paulo: CRE/PUC-SP e Edições Pulsar.

Buarque de Hollanda, S. (1995 [1936]) *Raízes do Brasil.* São Paulo: Companhia das Letras.

Capone, S. (2004) *A Busca da África no Candomblé: Tradição e Poder no Brasil.* Rio de Janeiro: Contra Capa Livraria/Pallas.

Chor Maio, M. and Monteiro, S. (2005) Tempos de raçialização: O caso da 'saúde da população negra' no Brasil. *História, Ciência, Manguinhos* 12, 419–446.

Costa, S. (2002) A construção sociológica da raça no Brasil. *Estudos Afro-Asiáticos* 24, 35–61.

da Cunha, E. (1973 [1902]) *O Sertões.* São Paulo: Cultrix.

Dantas, B.G. (1988) *Vovó Nagô e Papai Branco: Usos e Abusos da África no Brasil.* Rio de Janeiro: Ed. Graal.

FGV – Fundação Getúlio Vargas (2005) *Retrato das Religiões do Brasil.* São Paulo/Rio de Janeiro: Centro de Políticas Sociais, Instituto Brasileiro de Economia, FGV.

Freyre, G. (2003 [1933]) *Casa Grande e Senzala.* São Paulo: Ed. Global.

Governo do Estado, Secretaria da Cultura e Turismo (2005) *Livro do Século XXI – Consolidação do Turismo: Estratégica turística da Bahia 2003–2020.* Salvador, Bahia: A Secretaria.

Grünewald, R. de A. (2003) Turismo e etnicidade. *Horizontes Antropológicos 9,* 141–159.

Guimarães, A.S.A. (1999) Raça e os estudos de relações raciais no Brasil. *Novos Estudos CEBRAP* 54, 147–156.

IPHAN (2004) Parecer processo no. 01450.008675/2004-01. Salvador: Instituto de Patrimonio Histórico Artístico Nacional. On WWW at http://www.revista.iphan.gov.br/docs/parecer1.pdf. Accessed 23.2.07.

Lonely Planet (2005) *Brazil Travel Guide* (6th edn). Oakland, CA: Lonely Planet Publications.

Martínez-Echezábal, L. (1996) O culturalismo dos anos 30 no Brasil e na América Latina: Deslocamento retórico ou mudança conceitual? In M. Chor Maio and R. Ventura Santos (eds) *Raça, Ciência e Sociedade* (pp. 107–124). Rio de Janeiro: Editora Fiocruz.

Ministério da Saúde (2004) Seminário Nacional Saúde da População Negra, Brasília, agosto 2004. On WWW at http://bvsms.saude.gov.br/bvs/publicacoes/caderno_textos_basicos_snspn.pdf. Accessed 16.3.08.

Nina Rodrigues, R. (1899) Métissage, dégénérescence et crime. *Archives d'Anthropologie Criminelle* 14 (83), 477–516.

Nina Rodrigues, R. (1988 [1933]) *Os Africanos no Brasil.* São Paulo: Ed. Nacional.

Nogueira, O. (1955) Relações Raciais no Município de Itapetininga. In R. Bastide and F. Fernandes (eds) *Relações Raciais entre Negros e Brancos em São Paulo* (pp. 362–554). São Paulo: Anhembi.

Parés, L.N. (2006) *A Formação do Candomblé: História e Ritual da Nação Jeje na Bahia.* Campinas: Editora Unicamp.

Ramos, A. (1939) *As Coletividades Anormaes.* Rio de Janeiro: Ed. Civilização Brasileira.

Reinhardt, B. (2006) Espelho ante Espelho: A Troca e a Guerra entre o Neopentecostalismo e os Cultos Afro-Brasileiros em Salvador. Masters dissertation, Universidade de Brasília.

Romero, S. (1895) *O Evolucionismo e o Positivismo no Brasil.* Rio de Janeiro: Livraria Clássica de Álvares e C.

Roquete Pinto, E. (1927) *Seixos rolados.* Rio de Janeiro: Editora Mendonça, Machado & C.

Schwarcz, L.M. (1993) *O Espetáculo das Raças: Cientistas, Instituições e a Questão Racial no Brasil 1870–1930.* São Paulo: Companhia das Letras.

Skidmore, T.E. (1999) *Brazil: Five Centuries of Change.* Oxford: Oxford University Press.

Teles dos Santos, J. (2005) *O Poder da Cultura e a Cultura no Poder: A Disputa Simbólica da Herança Cultural Negra no Brasil.* Salvador, Bahia: Edufba.

Ventura Santos, R. (1996) Da Morfologia às Moléculas, de Raça a População: Trajetórias Conceituais em Antropologia Física no Século XX. In M. Chor Maio and R. Ventura Santos (eds) *Raça, Ciência e Sociedade* (pp. 125–139). Rio de Janeiro: Editora Fiocruz.

Wade, P. (1997) *Race and Ethnicity in Latin America.* London: Pluto Press.

Wang, Y. (2007) Customized authenticity begins at home. *Annals of Tourism Research* 34, 789–804.

Chapter 8

Tourism and the Making of Ethnic Citizenship in Belize

J. TERESA HOLMES

There is a large sign that stands by the side of the road at a busy intersection in downtown Belize City, the main coastal city of the Central American/Caribbean country of Belize. The sign reads 'Discourtesy is the Beginning of Violence': it is a 'reminder' sponsored, we are told, by a local law firm and the Belize City Council (see Figure 8.1). This sign is located on a major thoroughfare, used by many tourists as they make their way from the airport, the bus station or the water-taxi docks to the Fort George area of town, where the more expensive hotels are located. As a tourist, you see this sign and recognise that it is not directed at you. But you do understand, intuitively, that it has been erected because of your presence and because of the presence of the many other tourists who have been drawn to this small country by the government's active promotion of Belize as a tourist destination. Taken at face value, this sign obviously stands as a reminder to locals of the importance of courtesy in their daily behaviour. It also can be read as an indication of the significance placed on tourism and the tourism product by local government authorities. In fact, the sign and its message suggest that the expectations of proper public conduct and civic responsibility that are placed on Belizeans as citizens can be linked to the demands of tourism.

As you spend time in Belize, you encounter other indications of the importance of tourism in people's everyday lives. There are frequent announcements by the Belize Tourism Board on the local radio that remind Belizeans that 'tourism is for all of us'. Some of these radio spots echo the sentiments of the sign in downtown Belize City and remind Belizeans to 'be kind to tourists'. In yet another set of advertisements, various individuals explain how they derive satisfaction and a sense of fulfilment as a Belizean from their involvement in the tourism industry as a cook, a sculptor, a tour guide and a bartender. The proliferation of

Figure 8.1 Belize City street sign

these kinds of signs suggests that we should take seriously the ways in which state-based forms of tourism development are engaged in the production of citizenship or in the shaping of what it means to 'be Belizean'.

If we can trace a link between tourism and the production of citizenship in Belize, we can begin to ask what exactly is the role played by tourism in the construction of an idea of what it *means* to be a Belizean. How might tourism work to constitute proper civic conduct in Belize? We can begin to get answers to these questions by considering the increasingly common connection that is made, in the context of tourism development, between 'being Belizean' and being 'ethnic'. This is a new direction in Belize. The display and promotion of ethnic diversity has only been part of the government's official plans for tourism since 1998, but it has rapidly become an integral part of both marketing strategies for tourism and the promotion and conservation of forms of cultural heritage that are central to tourism development. In recent years, the government, along with private sector partners, has been promoting what are characterised as the distinct ethnic cultures of Belize in venues and at events that are sites of both international and domestic tourism.

These sites provide a setting in which state, regional and local authorities, as well as residents of host communities, exhibit and perform cultural forms that have come to be associated with what have been defined as the primary ethnic groups of Belize: Creole, Garifuna, Maya and Mestizo.[1] In these varying contexts, Belizeans are encouraged to display their ethnicity as tourist objects while, at the same time, engaging as tourists with the ethnicity of other Belizeans. Thus, through tourism development, as well as in other contexts, the government encourages local populations to recognise, display and live an ethnic identity as a civic responsibility and an expression of the national identity of Belize.

All of this suggests that in Belize we can trace the development of what Wood has called 'touristic ethnicity'. This refers to the 'incorporation of new and uniquely touristic modes of visualisation, experience and discourse into long-standing processes of cultural construction' (Wood, 1998: 224), in particular the construction of ethnicity. In his discussion of touristic ethnicity, Wood reminds us of the relational dynamics of ethnicity and points out that these dynamics include an inter-play between tourism and the state in the construction of ethnicity (Wood, 1998: 222).

Following from this observation, I want to explore the implications of what is, in Belize, a complex association between tourism and the state in which ethnicity is produced and managed as a condition of citizenship. But, here, the point is to rethink our understanding of the relational dynamics of ethnicity identified by Wood, by considering how the Belizean state, through the medium of tourism development, draws on and promotes notions of ethnicity that shape and make possible certain forms of citizenship. This approach recognises the state's ability to foster and manipulate cultural forms and cultural identities through directed strategies of tourism development, and it explores the politically salient aspects of this process.

This approach draws on the growing body of literature that, following the work of Foucault (1978, 1988) and Agamben (1998), examines the biopolitics of sovereign power, as exercised by the modern state (cf. Hansen & Stepputat, 2005; Ong, 2006). That is, how the very substance of life as 'bios' is transformed by the activities of the state into 'politically qualified life' (Agamben, 1998: 7). Here, I want to examine how, in the context of varying discursive strategies that currently shape tourism development, the Belizean state promotes ethnicity as the foundation for the 'sovereign subject' that is captured or expressed in the figure of the citizen (Agamben, 1998: 127–128). In this way, 'touristic ethnicity' is implicated in a state-sponsored refashioning of political being, as ethnicity is transformed into political substance, or into a condition of citizenship.

'Being Belizean': National Identity and Cultural Hybridity

If you pick up any guide book on Belize and look through the section on facts about the country, you will likely find reference to its ethnic diversity. For example, the first edition of the Lonely Planet guide to Belize notes that for 'such a tiny country... Belize enjoys a fabulous, improbable ethnic diversity' (Miller & Miller, 2002: 27). This guide, like most other sources, then goes on to describe the primary ethnic groups in Belize: Creoles, or 'the descendants of British pirates... and African slaves'; Mestizos or 'people of mixed Central American Indian and Spanish heritage'; the Yucatec, Mopan and Ke'kchi Maya; and the Garifuna or 'Black Caribs', who are described as being 'of South American Indian and African descent' (Miller & Miller, 2002: 27–28).

While ethnic diversity is often remarked upon as a unique character-istic of Belize in travel literature and in academic sources, until recently the government has downplayed the significance of differences in the contemporary ethnic identities of the Belizean people. After gaining full independence in 1981, and until quite recently, successive governments sought to create a 'Belizean' national identity that recognised the significance of ethnic diversity in the past, but that, at the same time, promoted an image of modern cultural hybridity. As Anne Sutherland notes, this portrayal 'focuses on historical and cultural differences [that come] together to form a unique Belizean identity' based upon 'a consciousness of a rich hybrid cultural *heritage*' (Sutherland, 1998: 78, my emphasis; see also Sutherland, 1996). In other words, for much of the last 25 years, the government of Belize has relied upon a notion of culture that relegates ethnic diversity to the past as a heritage that has shaped, but does not define, national identity.

This stance is very much in keeping with varying perspectives on national identity in many Caribbean and Latin American countries, where there is recognition of the cultural (and racial) heterogeneity of the population and a concomitant promotion of a national character that rests on the premise of a 'mixed' or emergent cultural whole. In Brazil, this can be seen in what Natasha Pravaz (2003: 120) has referred to as the 'myth of hybridity as national identity' that is envisioned as a process of 'whitening' and that is captured in the figure of the *mulatto*, or person of mixed racial descent (see also Calvo-González & Duccini, this volume). Similarly, in Mexico the discourse of *mestijage* (racial mixture) is a predominant theme in the nationalist narrative (Perez-Torres, 1998). The related concept of creolisation, the process by which diverse cultures and racial groups merge to create a new culture, is central to many national narratives in the Caribbean and is often played out in metaphors that stress a notion of 'mixture'. In Trinidad, for example, the dominant national narrative relies on the idea of a 'callaloo' (mixed or creole) society, understood, as Khan (2001: 284) notes, as 'that felicitous and mutually transforming mixing of cultural, racial, and religious diversity'. We can see the same underlying logic at work in Belize, where the term 'Creole' may be used specifically to refer to a population of Afro-European descent or more generally to 'refer to the blending of cultures into a national mixture' (Wilk, 2006: 108).

It is important to note, however, that these national narratives can be characterised by unresolved tensions. They celebrate diversity and, in the same moment, deny it through the promotion of a singular, emphatically authentic national identity. This tension is often effectively masked by the symbolic importance placed on a notion of national unity based on practices and concepts of liberal governmentality, discipline and other homogenising social and political forces. However, this does not mean that there are not other narratives of identity that run counter to that of a 'national mixture'. For example, Munasinghe (2002: 676) identifies two distinct nationalist discourses in Trinidad. One is the dominant callaloo narrative of the nation, espoused by a government associated with the Afro-Trinidadian population. The other, expressed most often by Indo-Trinidadians who are politically marginalised, relies instead on the metaphor of a 'tossed salad' to refer to a nation in which 'each diverse ingredient maintains its original identity'.

For much of its history as an independent country, the ethnic divisions within Belize have not been as marked, or as politically salient, as in other countries in the region. In a paper published in the mid-1990s, Wilk (1995: 113) points to the 'emergence of a public national culture' enthusiastically embraced by Belizeans and notes that community members, while not in complete agreement over the content of that national culture, emphasise the harmonious racial mixture that is Belize. At the same time, though, he argues that the presentation of ethnic diversity in such contexts as music, art, dance and 'custom' 'cordons off a potentially dangerous and divisive issue by placing it in the "safe zone" of expressive culture' (Wilk, 1995: 128). This points to those underlying tensions that, in the recent past, have been contained within, or contained by, the national narrative of a unified Belizean nation. This containment became increasingly important in the 1990s as ethnic conflicts, often expressed in discourses of ethnic difference, began to emerge due to the demographic shifts caused by an influx of Mestizo and Maya from Guatemala and Mexico and the efforts of varying ethnic groups to claim 'native' status (cf. Medina, 1999; Premdas, 1996). One could argue that what is now an emerging cultural tourism industry in Belize will continue to be a 'safe zone' for the expression of ethnic difference. However, it remains to be seen if the heightened, organised focus on ethnic diversity can contain the more politically charged assertions of ethnic identity in contemporary Belize.

In the past, the idea of a hybrid Belizean identity has been promoted most strongly by the People's United Party (PUP). The PUP became the dominant force in Belizean politics in the early 1950s, when the country, then known as British Honduras, was still a colony of the UK. In the late 1970s, the PUP organised 'across ethnic, class, and rural-urban lines, under a common banner of anti-colonial nationalism', mobilising a slim majority of the population to vote for independence in the 1979 elections (Merrill, 1992). When Belize became independent in 1981, the PUP formed the first national government. They have been the governing party for 17 of the past 27 years, including the period from 1998 to the recent elections in February 2008, when they were defeated by the United Democratic Party (UDP).

During the 1980s and 1990s, the PUP advocated a strategy of nation building that promoted the idea of 'being a Belizean' and encouraged 'various ethnic groups to work towards the *consolidation* of a community of interests of the "citizens"' (Roessingh & Bras, 2003: 166, emphasis added). For example, they endorsed certain events that were designated as national holidays, including Independence Day, St George Caye Day (which commemorates the defeat of the Spanish in 1878 by British colonists living on St George Caye) and Garifuna Settlement Day. While the latter two celebrations have special relevance for the Creole and Garifuna groups, they were encompassed within the government's nation-building project and treated, in government discourse, as symbols of national unity. Also, buildings associated with the British colonial past (such as Government House in Belize City), as well as architectural monuments associated with a Maya indigenous past, were called into service as national symbols and were said to reflect the shared cultural heritage of Belize.

One of the focal points of the PUP's earlier nation-building strategies was their attempt to build a national museum. Plans for the National Museum of Belize were first announced shortly after independence in 1981. The idea was to locate the museum in the newly built capital city of Belmopan, itself a symbol of national unity as it is populated by a cross-section of government workers and politicians originating from around the country. In 1992, the PUP government put forward specific plans for this museum, stating that it is 'the latent symbol of our new nation's urge for a "cultural renaissance"... a place where Belizeans will confront an accessible multidimensional exploration of their national history and identity' (Government of Belize, 1992).

The artist Joan Duran developed the plans for the structure and content of this museum, clearly articulating the ideological position of the PUP on national identity. In June 1993, two American anthropologists, Richard and Sally Price, visited Belmopan, met with Duran and were shown the plans for the museum. They report that Duran specifically, and rather vehemently, rejected the display of a series of collections focusing on the ethnic groups of Belize. He told the Prices that

> It's fine to know that people have different ways of doing things. Some people shoot each other from the back and others do it from the front. ...but it's counterproductive to constantly label people by ethnic identity. That kind of "tribalism" works against the national good... and this museum is being put together in the interests of tolerance and nation-building. (Price & Price, 1995: 98)

In keeping with this perspective, Duran's exhibits were to focus on images of the multi-faceted nature of Belizean society, displaying cultural heritage in such a way that it would symbolise 'the transcendent unity of the Belizean people' (Price & Price, 1995: 101). His vision was never realised. Shortly after the Prices visited Duran in 1993, elections were called and the PUP lost to the UDP. With the change in government, plans for the museum were shelved.

The PUP's proposal for a national museum coincided with their increasing focus on tourism development. During the 1980s, neither the PUP nor the UDP treated tourism as primary for national economic development. For example, while tourism was included as part of an overall strategy of diversifying the economy in the development plan for 1985–1989, agro-industrial development was the core economic strategy (see Government of Belize, 1984; drawn up by the government under the UDP). However, in the early 1990s, the PUP government officially formulated an 'integrated tourism strategy' that took into account 'the large potential of the tourism industry to contribute to economic development' (Ministry of Tourism and the Environment, 1994). It is important to note, though, that this tourism strategy did not encompass, or refer to, contemporary cultural forms. Instead, it relied on the images of 'untouched nature' and 'the Maya past' that had been central to a nascent Belizean tourism since the 1970s.

Given the government's efforts to construct a national museum in the early 1990s, it is significant that documents relating to economic development from this period did not mention the National Museum of Belize as part of an overall tourism strategy. In the official plans for the museum, it was stated that a 'major objective' is to 'attract tourism' (Government of Belize, 1992). However, this is mentioned only in passing and there is no consideration of how the museum could be promoted as a tourist attraction. This latter point would certainly have been a concern as the museum was to be located in Belmopan, a city that had no tourist amenities to speak of and was not a tourist destination or even a recommended stop on the tourist route. Thus, although the government's official plan was to 'promote Belizean eco-cultural tourism as a center-piece of its development marketing efforts', it appears that the only culture to be promoted and displayed as part of a tourism strategy was pre-historic, associated with Maya ruins (Ministry of Tourism and the Environment, 1994). At this time, the government's focus on eco-cultural tourism suggests that culture was to be treated as a natural resource, literally encased in the Belizean soil.

This approach to tourism development persisted throughout the 1990s, as when the UDP succeeded the PUP after the 1993 elections, they maintained a focus on eco-cultural tourism (Ministry of Economic Development, 1995). The official approach to tourism development began to change, though, after the PUP returned to government in 1998. Upon taking office, the party expressed a renewed commitment to tourism that now, for the first time, included the promotion of culture as a primary tourist attraction. Instead of re-affirming the importance of a unique, hybrid culture as they had in the early 1990s outside of the context of tourism, however, the PUP now began to celebrate the central role of ethnic diversity in tourism and in other realms of Belizean life.

This focus on culture, in the form of diverse ethnic cultures, remains important in Belizean tourism. Like many other countries of the southern hemisphere, Belize is responding to the demands of international tourism by making cultural tourism, and the promotion of regional culture and ethnicity in particular, a focal point for economic development (see Adams, 2006; Bruner, 2001, 2005: Chap. 8; Duval, 2005; Hill, 2007; Rutherford, 2003; also Calvo-González & Duccini, this volume). In Belize, this emphasis on ethnic diversity is mirrored in an emerging state (and tourism industry) discourse of multi-culturalism that brings to the surface a tension that was hidden in earlier government discourses of cultural hybridity. As noted earlier, the idea that Belize is characterised by a 'blending of cultures into a national mixture' (Wilk, 2006: 108) both recognises and subsumes diversity within the encompassing figure of the nation. However, this nation contains ethnic interests that increasingly diverge politically and economically. Now, as the state focuses on producing and managing populations so as to satisfy the demands of an international tourism industry for ethnic attractions, there is the potential to disrupt earlier notions of that national mixture and to redefine what it means to 'be Belizean'. Thus, it becomes possible to consider how, through this production of ethnicity, tourism policy and practice enacts a vision in which ethnic diversity is the very thing that creates and signifies a common national identity, or a sense of citizenship.

'Being Belizean': Ethnicity, Tourism and the State

In her ethnography on the responses of Belizeans to rapid globalisation, Sutherland notes that

> as Belize develops into a nation situated in the global ethnic community, there is increasing influence... of the idea that there is something 'natural', given, and ascriptive about ethnicity... a deeply felt shared sense of we-ness among people in a group who share the same race, language, kinship, gender, or place. (Sutherland, 1998: 80)

She goes on to argue that the emergence of an ethnic consciousness in Belize is a consequence of the 'ideoscapes about race, identity, and experience' that have been brought to Belize by European and North American tourists, the global media, international ethnic movements and Belizean émigrés (Sutherland, 1998: 80–81).

It is certainly true that such global forces introduce ideas about race, culture and identity that serve to fuel the production and display of ethnicity in Belize. Seeing these forces as external, however, presumes that Belizean culture is a bounded, unchanging unit that has been *acted upon* by the forces of globalisation (cf. Wood, 1998). This approach ignores the 'everydayness' of such forces in the lives of Belizeans and the possibility that tourism contributes to the shaping of culture in general, and ethnic culture in particular. This a process that is explored in the work of Picard (1996, 1997), who contends that, in what he calls 'touristic cultures', we see the emergence of a process of 'touristification' that proceeds from within by blurring the boundaries between the inside and outside, between what is 'ours' and what is 'theirs', between that which belongs to 'culture' and that which pertains to 'tourism' (Picard, 1997: 183; but see Theodossopoulos, this volume).

In his work on cultural tourism in Bali, Picard (1990: 74) has also shown that touristic culture is constructed as 'the product of dialogic construction between the Balinese and their various interlocutors, in a context defined by the growing integration of Bali within the overlapping networks of the internal tourist industry and the Indonesian state'. This highlights the significance of the relationship between tourism and the state in the process of 'touristification' and points to the inherent difficulties in trying to separate tourism from other social and cultural forces in many host communities. In recent years, a number of scholars have explored the connections between tourism and the state and have contributed significantly to our understanding of the ways in which the touristic production and display of ethnic culture is linked to such things as regional and national domestic politics and the symbolic interests of the nation (see, e.g. Adams, 2006; Bruner, 2005; Duval, 2005; Medina, 2003; Picard & Wood, 1997).

These issues are particularly relevant in the case of Belize, where tourism is central to recent governmental development projects, and thus is closely regulated. Certainly, it is useful to consider how the association between tourism and the state in Belize shapes the production and display of ethnic culture and, in turn, the construction of ethnic identity. However, I want to broaden the focus of analysis and ask a slightly different question. That is, in what ways might this emerging focus in Belize on touristic ethnicity reflect and shape aspects of a nationalist narrative of ethnic citizenship? What are the implications of this for the ways in which Belizeans manage and live their citizenship? To begin to explore these issues, it is necessary to consider the shift to culture, especially ethnic culture, as a directed government strategy for tourism in Belize.

One of the first indications that the government was taking a different approach to the role of culture in tourism was a simple name change. In the 1990s, the government body responsible for tourism development was called the Ministry of Tourism and the Environment. This was to be expected, given the focus on ecotourism (or eco-cultural tourism) in development strategies. After the PUP returned to government in 1998, the name was changed to the Ministry of Tourism and Culture. While this signalled a new perspective on the anticipated role of culture in tourism, more recent developments suggest that there has also been a change in the official perspective on what constitutes 'Belizean culture', and therefore Belizean identity. Increasingly, and especially in the context of tourism, the government's promotion of Belizean identity emphasises, rather than downplays, ethnic diversity.

This change in emphasis is also evident in the PUP's economic development strategy for 2003–2005. Here, for the first time in a government development plan, there is a lengthy section devoted to 'Tourism and Culture'. This document provides evidence of the government's commitment to culture, by noting the establishment of the National Institute of Culture and History, charged with the 'development, promotion, and preservation of Belize's culture' (Ministry of Economic Development, 2002: 36). One of the stated aims of this institute is the completion and opening of the Museum of Belize, not only in Belmopan, but also in the tourist entry point of Belize City. Unlike government discourse in the 1990s, this repository of Belize culture is now clearly tied to tourism development.

Significantly, in this same document (and also for the first time), one of the proposed tourism initiatives is to 'promote cultural tourism, focusing on the *rich diversity* of [Belizean] ethnic groups' (Ministry of Economic Development, 2002: 38, emphasis added). Among the primary venues for the promotion of this cultural diversity are the Houses of Culture. Four of these have been opened in what are generally considered to be distinct ethnic regions of Belize: in northern Belize in Orange Walk, where there is a large Mestizo population; in Benque Viejo, which is in western Belize on the border with Guatemala, in an area historically associated with Maya peoples; in Belize City on the central coast, associated primarily with a Creole population; and further south on the coast in Dangriga, a city identified very strongly with the Garifuna. These provide a good example of a venue that both produces touristic ethnicity and enables the construction of the Belizean citizen as 'ethnic'.

While the Houses of Culture, especially the one in Belize City, are marketed to international tourists as sites of cultural tourism, they are intended for the domestic tourism market as well. For example, at the opening of the Benque Viejo House of Culture in September 2001, the co-ordinator, David Ruiz, noted that this site would be 'the staging ground for exhibitions, cultural events and a place where the culture of the people will be preserved through educational activities designed for young people' (Office of the Prime Minister, 2001). These goals are pursued as Belizeans are encouraged to visit the various Houses of Culture to view cultural artefacts that primarily exhibit the ethnic culture of a specific region. At these venues, they are also encouraged to participate in cultural events, often conceived of as ethnic in origin, and engage in educational activities that, presumably, will produce a 'knowledge' of the distinct ethnic cultures of Belize.

The two most well-developed Houses of Culture are in Belize City and in Dangriga. They provide an example of how the government's goal of promoting cultural tourism might be achieved through a display of ethnic diversity that constructs an understanding of ethnic identity as central to what it means to be Belizean. The House of Culture Museum in Belize City is located in the former Government House, at one time the residence and the site of the administrative offices of the British Governors of Belize. Perhaps because it is located in the former seat of the colonial government and shares space with the National Institute of Culture and History, this particular House of Culture has retained some of the earlier focus on the idea of a contemporary and unified Belizean culture. In April 2005, the art gallery included exhibits that displayed a tension between the idea of hybrid culture and ethnic diversity, as a postmodern montage of street scenes in Belize City was placed next to a painting by the Garifuna artist, Pen Cayetano, depicting a ritual centring on the trickster figure of John Canoe, and a slate carving by the Garcia sisters whose cultural centre in western Belize is devoted to the preservation of the 'spirit of the ancient Maya' (Garcia Sisters, 2006). Yet, this tension is submerged, and the image of hybridity recedes, in the face of the celebration of ethnic culture that is fostered by this institution.

While the tension between cultural hybridity and cultural diversity is certainly present in this House of Culture, it is also contained rather neatly within the conceptual boundaries of the Belizean nation, as Government House remains one of the symbols of national unity. In the early years of independence in the 1980s, it symbolised a rejection of colonial domination and a national desire for independence and cultural unity. Today, this building no longer represents the transcending of a colonial past and the emergence of a unified Belizean population. Instead, it encompasses and objectifies a diverse cultural heritage, giving it a contemporary significance that rests in an uneasy tension with, but that also anchors, the more intangible forms associated with a hybrid culture.

The Gulisi Garifuna Museum, located on the outskirts of Dangriga on the main road into the city, is a joint venture of the National Institute of Culture and History and the National Garifuna Council (see Figure 8.2). While not an official House of Culture, it reflects the government's promotion of cultural tourism by focusing on the ethnic diversity of Belize. This museum is devoted solely to the presentation of the historical and contemporary culture of the Garifuna and thus demonstrates quite strikingly the emerging focus on distinct ethnic cultures in Belize. It is quite clear that this museum plays an important educational function for local Garifuna, as a repository of their cultural

Figure 8.2 The Gulisi Garifuna Museum, in Dangriga

heritage (see National Garifuna Council of Belize, n.d.). It is also promoted as a destination for international visitors staying at resorts and lodges along the coast. As such, it is an important venue in which a distinct sense of Garifuna identity is displayed and enacted for domestic and international tourist consumption. While providing a venue for the display of ethnicity that is similar to the House of Culture Museum in Belize City, the Gulisi Garifuna Museum, and the enthusiastic support that it enjoys from the local Garifuna population, also points to the important ways in which such sites can engage local populations in the production of citizenship.

The Belizean government's production and display of ethnic culture in sites such as the various Houses of Culture is particularly significant, given that Belizeans (at least those in official government positions) are engaged in a process of nation building. Since Belize gained independence, successive governments have sought to exercise sovereignty by forging a sense of national identity and citizenship and a place in the regional and global economy. In their recent discussion of postcolonial sovereignty, Hansen and Stepputat argue that the elites who take power in a newly independent nation devote

> enormous energy and pedagogical ingenuity to the task of converting
> colonial subjects into national citizens – capable of responsible public

conduct, loyal to the state and prepared to accept their responsibility as the backbone of society in return for privileges and recognition from the state. (Hansen & Stepputat, 2005: 26)

They also point out the 'production of citizenship', understood at least in part as the 'display of "civic conduct" in the public domain', takes place in 'various localities and contexts' (Hansen & Stepputat, 2005: 26). Following from their comments, I contend that tourism is one important context in which the production of citizenship occurs in Belize. The construction of touristic ethnicity in particular provides a venue for the regulation of civic conduct and thus is one means by which local Belizeans are managed and come to understand themselves as, above all, ethnic citizens of Belize.

In this way, the Houses of Culture are sites of domestic tourism that not only invite local residents to experience and understand their own ethnic culture, but also attract Belizeans from other areas of the country who are able to sample the ethnic diversity that is increasingly seen as a central element of Belizean society. What is more, these activities are specifically linked to what it means to 'be Belizean' and thus to act as a Belizean citizen. Perhaps one of the better examples of this linkage can be seen in a speech given by Said Musa, then Prime Minister, on the opening of the Benque Viejo House of Culture. In his statement to the assembled crowd, Musa remarked that

> our Belizean identity comes about as a result of our culture. . . . Culture is at the heart of our development. Ours is not a monoculture, but a diverse one where many groups are striving to make unity of this diversity to bring about the unique Belizean identity. (Office of the Prime Minister, 2001)

These remarks, and those of Ruiz that were described above, point to an assumption that, as this particular House of Culture (and by extension the other three) displays the 'diverse' cultures of Belize, it will also serve to further the goal of bringing about a 'unique Belizean [national] identity'. I argue that this is an indication of the complex association that exists in Belize between the production of touristic ethnicity and the production of citizenship. By 'exhibiting' and 'educating' Belizeans, the Houses of Culture (as well as other venues of touristic ethnicity) are seen as providing a setting in which domestic tourists are able to carry out their civic responsibility both by displaying and engaging with their own ethnicity and by developing a consciousness of the ethnicity of their fellow citizens.

Tourism and Citizenship: The Conduct of Conduct

My concern here has been to explore the discursive strategies utilised by the state as part of a political process that draws on and attempts to reshape cultural identity as a policy of tourism development. This suggests that there is an emerging and complex association between tourism and the state in Belize, through which ethnicity could be produced, managed and lived as a condition of citizenship. This is significant, given that recent literature on Belize has identified emerging political and economic tensions between local populations that are often expressed in terms of historical and cultural distinctiveness (Medina, 1999; Premdas, 1996). It raises the question of the larger political ramifications of tourist development policies that rely on the promotion, and manipulation, of essentialist forms of cultural identity. Belizean development policies promote the economic advantages, through tourism, of being ethnic, which draws attention to the processes of 'managing' and 'living' ethnicity. How could the recognition of these processes advance our understanding of the ways that individuals negotiate identity in the context of tourism, in response to the power of state-based manipulation of cultural forms and with respect to the divisive potential of politically charged concepts of ethnicity?

To begin thinking about these issues, I turn to another example of the centrality of ethnicity in Belizean tourism. In 2001, the Belize Tourism Board published a manual to be used in the National Tour Guide Training Program, which all tour guides must complete. This manual includes a large section that provides the prospective tour guide with information on the history and culture of Belize. It begins with the statement: 'Belize, "this beautiful jewel of ours", is a country of diverse ethnic/culture groups. ...Each has its own culture – although they sometimes overlap' (Iyo *et al.*, 2001: 203). It is interesting to note this shift from the earlier insistence that, despite diversity in heritage, contemporary Belizeans have a common, hybrid culture. Now it is the diversity that is marked and the hybridity that is mentioned in passing, almost as an afterthought. This highlights the increasing focus on ethnic diversity and suggests an emerging focus on ethnic difference as an aspect of 'being Belizean'.

In this training manual, one of the two roles outlined for the tour guide is as a 'host' who 'opens the doors of his/her home... to tourists, treating them as guests' and introducing them to this diverse culture (Christ & Palacio, 2001: 45). The manual suggests that this entails both an interpretation of ethnic culture and the particular kind of comportment that is associated with that culture. The manual states that 'as a good host, successful tour guides have the ability to entertain people with their stories and discussions and are expected to be courteous, patient, sensitive, caring, unselfish and even-tempered, yet firm with their guests' (Christ & Palacio, 2001: 45). This role of tour guide as host is significant, given that the National Tour Guide Training Program operates in the four districts of Belize, each seen as ethnically distinct. Since potential tour guides are generally trained in the programme located in their district,[2] their role would inevitably entail the performance, through stories and behaviour, of their own ethnicity. In other words, part of their role as tour guides, and representatives of Belize, would involve the management of their own ethnicity.

Thus, while the training of tour guides provides an example of the link between touristic ethnicity and civic responsibility, it also points to the 'managing' of one's ethnicity. Some of the implications of this managing are addressed in recent work on sovereignty and state power. Drawing on Foucault's (1978, 1983, 1991) notions of governmentality and bio-power, this work maintains that the creation of citizenship is linked to the 'production of a biopolitical body' (Agamben, 1998: 6; Hansen & Stepputat, 2005; cf. Ong, 1999). In a very general sense, the notion of biopolitics refers to a style of government (governmentality) that regulates populations. However, it also recognises the convergences of life and politics and suggests that, in the modern state in particular, political beings (citizens) are produced through the inclusion of some forms of 'life' and the exclusion of others. From this perspective, one could argue that 'life' is managed in Belize through the inclusion of 'ethnicity' as a way to live 'a good life' in which one is fully engaged as a citizen (Agamben, 1998).

In this sense, then, we could see tourism as a political act (cf. Cheong & Miller, 2000), a form of governmentality that seeks to define ethnic 'life' as an object of touristic consumption and thus to determine 'the conduct of conduct'. From the perspective of governmentality, you could understand touristic ethnicity as a form of what Foucault (1988: 17) referred to as the 'technologies of power' that 'determine the conduct of individuals'. In this regard, the tour guide training manual could be seen as a pedagogical tool that helps create responsible national citizens. Other tourist venues, such as the Houses of Culture, could also be seen as having a similar function, although perhaps in more muted form. And, certainly, the sign in downtown Belize City with its reminder of the importance of 'courtesy', like the incessant radio advertisements with their reminders to 'be nice to tourists', play a similar pedagogical and disciplinary role in Belizean society.

But this approach, by itself, neglects an important aspect of touristic ethnicity: how individual Belizeans might 'live' their ethnicity in the context of tourism. In his later work, Foucault began to consider how technologies of power interact with 'technologies of the self' and thus 'permit individuals to effect by their own means or with the help of others a certain number of operations on their own bodies and souls, thoughts, conduct and way of being' (Foucault, 1988: 17). Thus, this idea of the technologies of self refers to ways in which people manage their selves in society, but it also recognises that, in so doing, they are enabled or constrained by available discourses. This provides a useful context in which to begin thinking about how touristic ethnicity both acts to manage citizenship and provides a context in which to negotiate citizenship.

This is a crucial balance to keep in mind when considering the interplay between tourism and the state, or between the production of touristic ethnicity and the production of citizenship. While individual Belizeans are subject to discourses of ethnicity and civic conduct, they are not necessarily imprisoned by them. Recent research on tourism in Belize suggests that they can, and do, contest and remake notions of touristic ethnicity, as they visit tourist sites, interact with foreign tourists and perform ethnicity as tourist objects (cf. Medina, 2003). But the question remains, to what extent are people acting outside of received discourses of ethnicity? In what ways might Belizeans choose to work within or outside of essentialist notions of ethnic identity in the context of their participation in emerging forms of cultural tourism?

As a way of demonstrating the complexity of these issues, I will end with an anecdote that reveals the potentially troubling processes involved in 'managing' and 'living' ethnicity as a form of citizenship in Belize. During a field trip in 2005, my research partner and I hired a tour guide, Luis (a pseudonym), to take us through Xunanatunich and Caracol, two of the more popular archaeological tourist sites in Cayo District, in western Belize, referred to in government and tourist discourse as a 'Maya' region. Luis came from the small town of San Jose Succotz, identified in government and tourism literature as having a long Mayan heritage. On visits to Xunantunich and Caracol, Luis provided what were clearly well-rehearsed accounts of the history and culture of the Maya people who lived in these places in the past. But, in moments of unscripted conversation, one thing that he made very clear was his own sense of ambivalence towards his relationship with these ancient Maya. When describing some aspect of the Maya past, for example, he would say that 'we don't do that anymore' or that 'to be modern' is not to be Maya.

On one particular trip to Caracol, though, he began telling us about the ghosts that he had seen there and about the footsteps and disembodied voices that he and others had heard coming from the ruins. He told us that he knew that the ghosts were the Maya ancestors of Caracol, but he was not sure what the voices were saying because he could not make out the language. He had been told that the voices spoke an ancient Maya dialect, but he personally did not know anyone who would know what the ghosts were saying. At this point, Luis's disclaimers about his Maya-ness were tinged with a sense of loss and a sense of dread at his inability to connect with an ethnic past that is still palpable, as it is so central to government discourses of tourism.

In other circumstances, as when he rejected an identity as 'Maya', it appears that he was attempting to play out the theme of ethnicity by 'playing out of' the box of *primordial* ethnic identity that has come to characterise notions of touristic ethnicity in Belize (cf. Sutherland, 1998: 80). However, his stories of the ghosts of Caracol demonstrated an awareness of the continuing significance of a way of being, or a way of being 'ethnic', that still haunts the world of the living. As such, Luis's ambivalence towards claiming an ethnic identity, and his simultaneous unease or sense of dread at his inability to enact ethnicity, is an eloquent illustration of the ongoing attempts of contemporary Belizeans to negotiate an emerging form of ethnic citizenship, and thus to come to terms with what it means to be Belizean.

Notes

1. Other groups are often identified as contributing to the ethnic makeup of Belize. These include: Lebanese and Chinese populations who migrated to Belize beginning in the middle of the 19th century and took up occupations as shopkeepers and traders; East Indians who came in the latter part of the 19th century as indentured labour; and Mennonites who arrived in the middle of the 20th century and are important in agriculture (Sutherland, 1998: 24–25). While these groups may be mentioned in tourism literature or websites, they are not promoted as a tourist attraction, perhaps because their ethnicity is not understood as authentic in a Central American or Caribbean context.
2. Personal communication, Dino Jimenez, March 2005.

References

Adams, K.M. (2006) *Art as Politics: Re-Crafting Identities, Tourism, and Power in Tana Toraja, Indonesia.* Honolulu: University of Hawai'i Press.

Agamben, G. (1998) *Homo Sacer: Sovereign Power and Bare Life* (D. Heller-Roazen, trans.). Stanford, CA: Stanford University Press.

Bruner, E. (2001) The Maasai and the Lion King: Authenticity, nationalism, and globalization in African tourism. *American Ethnologist* 28, 881–908.

Bruner, E. (2005) *Culture on Tour: Ethnographies of Travel.* Chicago, IL: University of Chicago Press.

Cheong, S-M. and Miller, M.L. (2000) Power and tourism: A Foucauldian observation. *Annals of Tourism Research* 27, 371–390.

Christ, S. and Palacio, V. (2001) Chapter one: Belize today. In S. Christ, S. Matus and V. Palacio (eds) *National Tour Guide Training Program: Trainers Manual* (pp. 1–41). Belize City: Belize Tourism Board.

Duval, D.T. (2005) Cultural tourism in postcolonial environments: Negotiating histories, ethnicities, and authenticities in St Vincent, East Caribbean. In C.M. Hall and H. Tucker (eds) *Tourism and Postcolonialism: Contested Discourses, Identities and Representations* (pp. 57–75). London: Routledge.

Foucault, M. (1978) *The History of Sexuality* (Vol. 1). New York: Pantheon Books.

Foucault, M. (1983) The subject and power. In H.L. Dreyfus and P. Rabinow (eds) *Michel Foucault: Beyond Structuralism and Hermeneutics* (pp. 208–228). Chicago, IL: University of Chicago Press.

Foucault, M. (1988) Technologies of the self. In L.H. Martin (ed.) *Technologies of the Self: A Seminar with Michel Foucault* (pp. 16–49). London: Tavistock.

Foucault, M. (1991) Governmentality. In G. Burchell, C. Gordon and P. Miller (eds) *The Foucault Effect: Studies in Governmentality* (pp. 87–104). Chicago, IL: University of Chicago Press.

Garcia Sisters (2006) Spirit of the Ancient Maya – Tanah. On WWW at http://www.awrem.com/tanah/.

Government of Belize (1984) *Five-Year Development Plan, 1985–1989: Towards a Development Strategy – Draft Outline.* Belmopan: Government of Belize (Belize Archives, MC1331).

Government of Belize (1992) *The National Institute of Culture and History.* Belmopan: Government of Belize (Belize Archives, MC4824).

Hansen, T.B. and Stepputat, F. (eds) (2005) *Sovereign Bodies: Citizens, Migrants, and States in the Postcolonial World.* Princeton, NJ: Princeton University Press.

Hill, M.D. (2007) Contesting patrimony: Cusco's mystical tourist industry and the politics of *Incanismo. Ethnos* 72, 433–460.

Iyo, J., Palacio, J., Palacio, V. and Ruiz, C. (2001) Chapter four: History and culture of Belize. In S. Christ, S. Matus and V. Palacio (eds) *National Tour Guide Training Program: Trainers Manual* (pp. 203–251). Belize City: Belize Tourism Board.

Khan, A. (2001) Journey to the center of the earth: The Caribbean as master symbol. *Cultural Anthropology* 16, 271–302.

Medina, L.K. (1999) History, culture, and place-making: 'Native' status and Maya identity in Belize. *Journal of Latin American Anthropology* 4 (1), 133–165.

Medina, L.K. (2003) Commoditizing culture: Tourism and Maya identity. *Annals of Tourism Research* 30: 353–368.

Merrill, T. (ed.) (1992) *Belize: A Country Study.* Washington, DC: GPO for the Library of Congress. On WWW at http://countrystudies.us/belize.

Miller, C. and Miller, D.C. (2002) *Lonely Planet Belize.* Oakland, CA: Lonely Planet.

Ministry of Economic Development (1995[?]) *The National Development Strategy (Draft) 1996–2000.* Belmopan: Ministry of Economic Development, Government of Belize (Belize Archives, MC4493).

Ministry of Economic Development (2002) *Medium-term Economic Strategy 2003–2005.* Belmopan: Ministry of Economic Development, Government of Belize (Belize Archives, MC4493).

Ministry of Tourism and the Environment (1994) *Five-Year Plan, 1994–1998.* Belmopan: Ministry of Tourism and the Environment, Government of Belize (Belize Archives, MC4534).

Munasinghe, V. (2002) Nationalism in hybrid spaces: The production of impurity out of purity. *American Ethnologist* 29, 663–692.

National Garifuna Council of Belize (n.d.) National Garifuna Council of Belize. On WWW at http://www.ngcbelize.org/content/view/3/1/.

Office of the Prime Minister (2001) Benque Viejo Town Gets House of Culture. Release from Press Office. On WWW at http://www.governmentofbelize. gov.bz/press_release_details.php?pr_id = 1075.

Ong, A. (1999) *Flexible Citizenship: The Cultural Logics of Transnationality.* Durham, NC: Duke University Press.

Ong, A. (2006) *Neoliberalism as Exception: Mutations in Citizenship and Sovereignty.* Durham, NC: Duke University Press.

Perez-Torres, R. (1998) Chicano ethnicity, cultural hybridity, and the Mestizo voice. *American Literature* 70 (1), 153–176.

Picard, M. (1990) Cultural tourism in Bali: Cultural performances as tourist attraction. *Indonesia* 49, 37–74.

Picard, M. (1996) *Bali: Cultural Tourism and Touristic Culture.* Singapore: Archipelago Press.

Picard, M. (1997) Cultural tourism, nation-building, and regional culture: The making of a Balinese identity. In M. Picard and R.E. Wood (eds) *Tourism, Ethnicity, and the State in Asian and Pacific Societies* (pp. 181–214). Honolulu, HI: University of Hawai'i Press.

Picard, M. and Wood, R.E. (eds) (1997) *Tourism, Ethnicity, and the State in Asian and Pacific Societies.* Honolulu, HI: University of Hawai'i Press.

Pravaz, N. (2003) Brazilian *Mulatice*: Performing race, gender, and the nation. *Journal of Latin American Anthropology* 8 (1), 116–147.

Premdas, R.R. (1996) *Ethnicity and Identity in the Caribbean: Decentering a Myth.* (Working paper 234.) Notre Dame: The Kellogg Institute, University of Notre Dame.

Price, R. and Price, S. (1995) Executing culture: Musée, museo, museum. *American Anthropologist* 97, 97–109.

Roessingh, C. and Bras, K. (2003) Garifuna Settlement Day: Tourism attraction, national celebration day, or manifestation of ethnic identity? *Tourism, Culture and Communication* 4, 163–172.

Rutherford, D. (2003) *Raiding the Land of the Foreigners: The Limits of the Nation on an Indonesian Frontier.* Princeton, NJ: Princeton University Press.

Sutherland, A. (1996) Tourism and the human mosaic in Belize. *Urban Anthropology* 25, 259–281.

Sutherland, A. (1998) *The Making of Belize: Globalization in the Margins*. Westport, CT: Bergin and Garvey.

Wilk, R. (1995) Learning to be local in Belize: Global systems of common difference. In D. Miller (ed.) *Worlds Apart: Modernity Through the Prism of the Local* (pp. 110–133). London: Routledge.

Wilk, R. (2006) *Home Cooking in the Global Village: Caribbean Food From Buccaneers to Ecotourists*. Oxford: Berg.

Wood, R.E. (1998) Touristic ethnicity: A brief itinerary. *Ethnic and Racial Studies* 21, 218–241.

Chapter 9
Tourism and its Others: Tourists, Traders and Fishers in Jamaica

GUNILLA SOMMER and JAMES G. CARRIER

Recently, Mimi Sheller and John Urry (2004: 6) called for a 'de-centring of tourist studies away from tourists'. In this chapter, we heed their call. We do so by looking at something that is often ignored in anthropological work on tourism, the tourism sector, by which we mean the firms, employees and relations that are intended to cater to and profit from tourists. Our concern is with the ways that the people and firms in that sector seek to secure the sector's collective interest in the success of tourism in the face of other sets of people with other interests in the tourist destination.[1]

Our focus is how people's understanding of the world, expressed in their statements about it, can secure their interests and authority in relation to others. Implicit in this, of course, is the point that those understandings are not a neutral reflection of the world. Rather, they are necessarily partial, reflecting their particular perspectives and concerns. We will use this approach to make sense of some of the ways that people in tourism in the Jamaican tourist town of Negril talk about aspects of their world. More narrowly, we will describe how those in the tourism sector talk about different sets of people in ways that truncate or ignore relationships between themselves and those people, and so fetishise them (e.g. Strathern, 1996; classically Marx, 1867).

By this, we mean that the knowledge that people in the tourism sector have and put forward about, for instance, tourists, can ignore the context in which tourists exist, a context that includes the tourism sector. Ignoring the relationship between tourists and the sector fetishises tourists because it turns them into a set of distinct people who confront and constrain the sector. Doing this makes it probable that the attributes of tourists, to continue the example, that are shaped by the sector will be seen as inherent in tourists, which is what Pierre Bourdieu (1977) called 'misrecognition'. Whether intentional or not, this fetishism and its associated misrecognition are likely to have a political dimension. The world that they portray can make it seem obvious that credit for what is good or that blame for what is bad should be assigned in certain ways rather than others, that certain courses of action should be followed rather than others, and hence that certain interests should be protected rather than others.

Negril Tourism

For over a hundred years, efforts have been made to attract tourists to Jamaica (F. Taylor, 1993). Such efforts were not very successful until the 1950s, when the port of Montego Bay, on the north-west coast, became a tourist destination. It is the largest destination on the island and its airport handled three-quarters of the 1.35 million tourists entering the country in 2003 (Bakker & Phillip, 2005: Fig. 18). The town of Negril is about 50 miles (80 km) to the west of Montego Bay, at the western tip of Jamaica. When Montego Bay tourism began to flourish, Negril was only a small settlement and fishing beach that was not to have electricity until 1963. The nearest town was 18 miles (29 km) to the south-east, Savanna-la-Mar; people coming from Montego Bay commonly travelled by boat, wading ashore because there was no wharf.

During the 1960s, Negril began to attract a few American visitors who enjoyed the isolation and, it is said, access to marijuana and freedom from the American military draft. In spite of their presence, and the connection of Negril to the coastal road network, in 1971 it remained the case that 'most residents were higglers, farmers and fishermen' (Olsen, 1997: 286). Early in the 1970s, visitors increased, and some built houses on what is now called the West End, an area with dramatic coastal cliffs (Figure 9.1). According to some Americans long resident in Negril, a handful of these people built accommodation to house visiting friends. These became the first hotels, often with accommodation in a few individual cabins rather than in a common building, and some survive under different owners.

As the 1970s passed, Negril began to change, partly as word of the place spread and partly as a result of government policy. Gradually, hotels began to be built along the Norman Manley Boulevard, which had been completed in 1959. It led away from the West End along the aptly named Long Bay, with its stunning seven-mile beach. It divided the beach from the Great Morass, a large marsh on its in-land side. These early hotels were fairly small and conventional, mostly on the beach side of the boulevard, run by their owners.

Figure 9.1 The West End cliffs at Negril

However, the sector expanded and changed rapidly. The first all-inclusive hotel in Jamaica (Negril Beach Village, now Hedonism II) appeared on the Boulevard in 1976. As the 1980s turned into the 1990s, the hotels being built were further from the town, larger and more likely to be all-inclusives run by Jamaican and foreign corporations. Beginning in the 1990s, some of the earlier hotels became all-inclusive as well. These trends continue, with new hotels that are so self-contained and far from town that their relationship with Negril is more a matter of advertising than of geography.

The result of this development is a town with a population of about 10,000 that is the country's third largest tourist destination. It has about 100 hotels, the largest with 420 rooms. Between them, they have about 5850 rooms, almost a quarter of the country's tourist accommodation, about 60% of which are in all-inclusive hotels. In 2003, the place attracted about 275,000 tourists and its hotels employed almost 7800 people (Bakker & Phillip, 2005). This is the Negril tourism sector we describe in this chapter.

What we say about it is based primarily on research carried out by one of us (GS) from November 2004 to February 2005. This is complemented by what we learnt in the project of which that research is a part (overseen by JGC), which ran from January 2004 to June 2005. That project was concerned with the ways that fishers, conservationists and those in tourism are related to and understand the coastal waters, the people who use them and efforts to protect them. The study of the tourism sector involved informal observation of and semi-structured interviews with about 70 people, ranging from beach boys to present and former government ministry staff. However, the bulk of those interviewed were hotel managers, owners and environmental officers, especially at all-inclusive beach-front hotels. Additionally, there was informal observation at trade meetings, and informal interaction and interviews with tourists at hotels and beaches.

We said that we are concerned with the ways that those in the tourism sector construe their world. Before looking at those constructions, we will sketch the ambiguous position of tourism in the country. This will provide background for our main focus, the ways that those in the sector portray in a fetishised way some of the sets of people who are significant for that sector. Those sets of people are the tourists, traders and fishers of our title.

The Problem with Tourism

Tourism is important for Jamaica's economy. The country's conventional agricultural exports have been declining for decades, and it underwent structural adjustment in the 1980s (Bartilow, 1997; Bloom *et al.*, 2001; Witter, 2005: 191–194), which weakened its domestic economy and increased its imports. As a result, tourism has become more important for the economy generally, and especially as a source of foreign exchange. Between 1999 and 2003, the number of tourists varied between 1.25 and 1.35 million per year (unless otherwise attributed, statistics and statements about government policy in this paragraph are from Bakker and Phillip, 2005). They spent between US$1.25 and US$1.35 billion, which made tourism 'the largest source of foreign inflows outside of private capital' (i.e. remittances) (O'Neil, 2003: 4). In 2003, tourist expenditure amounted to 11.8% of the country's gross domestic product, a figure the government intended to raise to 15% by 2010. This tourism is exemplified by tourist hotels, which take almost 60% of tourist expenditure and employ about 30,500 people. In 2003, there were about 24,600 rooms in the country to accommodate tourists, of which 41% were in all-inclusive hotels and 28% in other hotels, and the government planned to add 12,000 rooms by 2010.

Those hotels are the most visible part of tourism, and are the focus of much of the criticism of the sector. That criticism is sufficiently pervasive that the first page of the government's tourism master plan (Jamaica, 2003: 1) states: 'Some people see tourism as an industry that benefits only a chosen few. Tourism needs to be seen as an industry that benefits everyone and the country as a whole'.[2] The basis of that criticism is perhaps epitomised in the image of the hotel beach. There, foreign guests occupy a lovely location that requires substantial resources to maintain. The only Jamaicans there are waiters and security guards, the former to serve tourists imported foods and the latter to make sure that no ordinary person penetrates this special space.

The criticisms that concern us are environmental and socio-economic, and they are intertwined. Critics say that tourism uses disproportionate quantities of resources, degrades the coastal environment and is unjust socially and economically. These criticisms are hardly unique to Negril and Jamaica, where they are portrayed in Stephanie Black's (2001) film *Life and Debt*. Throughout the region (e.g. Grandoit, 2005; McLaren, 1998; Panos Institute, 1998; Pattullo, 1996), and more generally in poorer countries (e.g. L. Burke *et al.*, 2000; Davies & Cahill, 2000; UNEP, 2002), tourism is associated with

> socioeconomic inequality and spatial unevenness; . . . environmental destruction. . . and rising alienation among the local population because of problems such as increasing crime, overcrowding and overloaded infrastructures, pollution and other environmental damage, conflicts over access to scarce resources. (Brohman, 1996: 53–54; see also Boissevain & Selwyn, 2004; M. Crick, 1989)

Of course, the sector needs just those things that critics say it harms. It needs the stable beaches, clear water, fish and coral that it advertises to attract its guests, rather than dying coral and declining fish stocks (e.g. Aiken & Haughton, 1987: 139–143; Haley & Clayton, 2003). Yet critics assert that 'every little thing that is done to accommodate tourists sets Jamaica one step back on the environmental scale' (Kozyr, 2000; see also Island Resources Foundation, 1996; Jamaica Environment Trust and Northern Jamaica Conservation Association, 2005; von Maffei, 2000). For instance, hotels are often constructed on unstable beaches; many Negril hotels have encroached on the Great Morass and one large hotel was built in a fish nursery area banned to fishers. In addition, tourist hotels produce substantial effluent: in 1994, three-fifths of the hotel waste water in Jamaica was either untreated or treated inadequately (R. Burke, 2005: 11); the average tourist in Jamaica produces almost four times as much solid waste daily as the average resident (Thomas-Hope & Jardine-Comrie, 2005: 3); tourist pleasure boats and jet skis pollute the waters. Tourists use disproportionate amounts of scarce resources, especially water: in Negril, five times as much per capita as residents (Olsen, 1997: 288). And when money is spent on basic infrastructure, it goes to tourist areas rather than parts of town where people live.

The sector also needs cheerful and supportive Jamaicans rather than disaffected and dispossessed locals, just as it wants to avoid the 'social unrest [that is] related to the inequitable distribution of tourism's benefits', exacerbated by the domination of all-inclusive hotels (Blackstone Corporation, 2001: 2; see also Eversley, 2003). This inequitable distribution includes the reluctance of all-inclusive hotels to buy food from local sources rather than import it from the USA, long a concern throughout the region (Bélisle, 1983). It also includes the apparent willingness to exploit local labour: alone among substantial sectors of the economy, tourism tends to pay its workers less than the minimum wage, about US$30 per week in 1999 (CIDA, 2002). Some hotel workers are paid even less: a Negril local government official received reliable reports that one large all-inclusive hotel was hiring staff on the condition that they receive no pay for their first six months (on labour law in tourism, see O. Taylor, 2002). Also, the sector is criticised for the unacceptable behaviour of tourists, ranging from rudeness toward local people and morally lax dress and behaviour in public places, to sex tourism (Caribbean Association for Feminist Research and Action, 2002; but see Pruitt & LaFont, 1995).

To help protect itself from such criticisms, the sector has supported efforts to improve things. The most obvious and institutionalised of these concern the environment, in particular two international forms of certification, the Green Globe (www.greenglobe.org) for hotels and the Blue Flag (www.blueflag.org) for beaches. These are awarded primarily on the basis of the environmental practices, educational activities and condition of the hotel or beach concerned. At the time of research, 29 Jamaican hotels had received the Green Globe. There were just four Blue Flag beaches, but the programme was only in its first year of operation in the country and several more beaches were working towards certification.

These certifications are about the natural environment. However, often those in tourism present them, and appear to understand them, as part of a larger set of policies and practices that benefit Jamaicans. Thus, when asked about the conservation practices of their hotels, people routinely and spontaneously talked as well about hotel projects that help local people. The most common of these was supporting local schools by helping fund the building or maintenance of classrooms, or by providing pupils with materials, meals and the like. Other projects ranged from providing free hotel accommodation for a group of foreign tourist dentists in return for their offering free dental care for needy local people, through contributing food for beach clean-up days, to supporting a community library or funding poverty-alleviating projects in Kingston, the country's capital. Beyond describing these sorts of specific projects, people often stressed the many jobs that their business created, and how this helped alleviate poverty.

However, there is more involved in securing the position of hotels against criticism than announcing the ways that a hotel company is being helpful. There is also an understanding of the world that sees the sector as constrained in its ability to do more by forces beyond its control, some of which are more damaging than the sector. This understanding does not represent the sector as heroic, but as doing all that can reasonably be expected. The forces beyond the sector's control are associated with different sets of people who affect tourism in Negril and who are rendered in fetishised and stylised form. Conversations with people in tourism in Negril readily turn to these sets of people. Important among them are government, commonly seen as incompetent and corrupt, and international bodies, commonly seen as ignorant of the realities of Jamaica and overly bureaucratic.

Also important are the sets of people that concern us, tourists, traders and fishers. The ways that those in tourism represent and invoke these sets are problematic. This is not because they are constructed in ways that have no relationship to the people they portray: they are stylised representations, not fabrications. Rather, what makes them problematic is that they are represented as being external to and independent of the tourism sector. As we shall argue, they are not that independent, for they are linked with and shaped by the sector. This means that these representations are, as we described earlier, fetishistic: they divide up the world in ways that ignore how the practices and policies of the sector are implicated in the existence and attributes of these sets of people.

By 'tourists', we mean guests at all-inclusive beach-front hotels (see Figure 9.2). These are not the only tourists in Negril. However, they are the majority of tourists, the firms that run those hotels are the most powerful, and they predominate in discussions of tourism in Negril and throughout Jamaica. By 'traders' we mean the local petty entrepreneurs who seek to deal with tourists. These range from taxi drivers, food vendors and hair braiders, to prostitutes and drug sellers. By 'fishers', we mean the body of Jamaicans who are in-shore, artisanal fishers using Negril's waters, most of whom use the Negril town fishing beach, located close to the road junction that is the centre of town (the project of which

Figure 9.2 An all-inclusive beach-front hotel on Long Bay

this research is a part included a year's field work among Negril fishers, allowing us to assess the accuracy of this rendering of them).

These sets of people are significant in tourism in Negril, though in different ways. Key among them is tourists, for two reasons. First, their relationships with the other two lie at the heart of the sector's interpretation and representation of the world. Second, the sector claims a special competence in understanding tourists: it knows what they want, and survives only by satisfying that want. We begin by describing how those in the sector present tourists and their relationship with traders. We then turn to their presentation of tourists and their relationship with fishers.

Tourism, Tourists and Traders

Those in tourism see the sector as beneficial and would like to incorporate more local people in it, to make the country more attractive to tourists and to make tourism more attractive to the country. So, for instance, there are sporadic efforts to have hotel guests meet ordinary Jamaican families and the Jamaican Tourist Board operates a Meet-the-People programme, and in marketing materials Jamaicans are presented as warm, attentive, happy and friendly (A. Crick, 2002; Duperly-Pinks, 2002; for the Caribbean generally, see Gmelch, 2003; Sheller, 2004). Quite a few of the tourists interviewed seemed to think the same, saying that they wanted to experience the 'real Jamaica'[3] not included in the standard tours, by talking to local people and going to local places (access to this 'real Jamaica' appears to be one reason for female tourists allying themselves with Jamaican men: Pruitt & LaFont, 1995).

However, those in the sector say that they are constrained by the fact that so many of the Jamaicans that tourists meet are not very pleasant. The archetype of the unpleasant Jamaican is the street trader. Unlike the sort of people in Kandy whom Malcolm Crick (1994: Chap. 5) describes, who seek to guide tourists to particular restaurants or shops, for instance, for a commission from the owners, these Jamaican traders approach passing tourists and try to sell things, at times insistently. These are the people who the Prime Minister, in a budget debate in 1985, referred to as 'touts, pimps, hustlers and drug pushers' (in Chambers & Airey, 2001: 111). Their behaviour is called 'hassling' or 'harassing' tourists, and people are regularly told that this behaviour harms Jamaica's reputation as a tourist destination, just as Crick says happened in Kandy.

> The anti-harassment programme initiated in 1996 and continued into 1997 was not successful during the year in stemming this scourge on the industry. As tourist harassment continues and escalates, the problem is being regarded as a serious deterrent to growth in the local industry. (PSOJ, 1997: 72; also PSOJ, 1999: 6. See also A. Crick, 2002; Duperly-Pinks, 2002; Jamaica, 2003)

A water sports manager in an all-inclusive hotel in Negril likened traders to 'Tasmanian devils. They look like angels, but when you turn your back to them they are devils'. Concerns in the sector about the risk of unpleasant contact with locals is manifest in practices that make it difficult for hotel guests to interact with Jamaicans other than hotel staff and contractors, plus the occasional taxi driver. Guests are warned against going outside resort areas because of the risk of unpleasantness, and they are assured that all their needs can be met within the resort. (Some tourists go to Negril hotels precisely for the sex and drugs available outside of them. The town has a reputation for this sort of thing, and the hotels profit from their business. Interviews indicate that the sector makes no sustained effort to eliminate this sort of activity, those in tourism arguing that attempts to do so are fruitless.)

This construction of the trader helps protect the tourism sector from critics who complain that the hotels keep their guests in enclaves. Those in tourism say that they cannot encourage greater contact because doing so would expose tourists to the unacceptable behaviour of many of the

Jamaicans that they meet on the streets. Underlying such an assertion is the implicit claim that the sector confronts two sets of people who are independent of it, and that the attributes of these two sets are incompatible. One set is traders, with their avarice; the other is tourists, with their self-indulgent desires.

This implicit claim is clearest with regard to tourists, who are portrayed as consumers making independent choices. This portrayal is illustrated in an interview with Gordon ('Butch') Stewart, who runs Sandals, a regional chain of all-inclusive hotels based in Jamaica. After the interviewer said that some object that all-inclusive hotels isolate tourists from Jamaicans, Stewart replied: 'The consumer is king' (European Commission, 1999–2000). Stressing consumer choice conceptually separates tourists, with their desires, from the sector, with its facilities. Such a separation fetishises tourists, for it ignores the relationship between tourists and the sector, one that shapes the desire that guides tourist choice.

Shaping tourist desire begins with the ways that advertisements portray Jamaica, as 'Your personal paradise', 'A little piece of heaven' or as 'A dream come true', what Echtner and Prasad (2003: 672–675) call 'the myth of the unrestrained'. The pictures accompanying these words are the 'iridescent, turquoise-blue Caribbean' and its white beaches, with beautiful, smiling, sun-tanned, white couples playing in the water, relaxing on sun beds or having colourful drinks (such images are hardly exclusive to Jamaican tourism; see Dann, 1996). The self-indulgence is justified by telling the potential tourist, 'You deserve the best' or 'You deserve to be pampered'.

In saying that this imagery shapes tourists' desires, we are not saying that it causes them. People do not carefully evaluate the images any more than they expect the reality to conform fully to them. Equally, we are not saying that tourists are anxious to resist the image by, for instance, seeking out extensive contact with local people. After all, as Malcolm Crick (1989: 327) asks, if going on holiday is just an unchanged continuation of life at home, then 'Why go at all?' The power of these images is their promise 'to change the order of everydayness' (Wang, 2000: 169). Or, as several of those in tourism put it, 'People come to get away. They just want fun in the sun and the water... [they] don't want to know in depth' about Jamaica.

To say that tourists are attracted by these images is not, however, to say that their desires need take the form that those in the sector say that they do. Before they travel, people could be exposed to images of self-indulgence that create expectations not so constrained by the hotel's perimeter fence. These need not entail knowledge and contact 'in depth', but more than the sector ascribes to tourists. These are the sets of images that Dann (1996) calls 'Paradise controlled' and 'Paradise confused', and images portraying greater contact were used in Jamaican advertising late in the 1970s (Bolles, 1992: 32). In addition, once tourists arrive, the sector is able to refine their preferences and actions. After all, tourists are away from home and likely to lack the sort of knowledge that would guide them in more familiar surroundings. They are, then, prone to do what their hotels suggest, which need not mean restricting themselves to in-house food, activities, entertainment, souvenirs and tour desks (see Cheong & Miller, 2000: 380–382).

Thus, through the expectations conveyed in their advertising and through their policies that restrict contact with local people, the sector implicitly encourages ignorance and even fear of Jamaican people. This fearful ignorance, which helps keep tourists in their hotel compounds, was manifest by many of the tourists that Sommer interviewed. They were shocked to learn that she had walked on her own in downtown Kingston and come back with all her belongings, that she had eaten food from street vendors and had got around using the local route taxis rather than the charter taxis that hotels arrange. These are the sort of activities that hotels warn tourists against.

Equally, hotel policies influence the behaviour of traders and their interaction with tourists. Recurrent pressure on local authorities from the sector to clear unlicensed traders from tourist areas increases traders' hostility toward the sector. Similarly, restricting contact between tourists and Jamaicans reduces the chance for traders to learn about tourists, just as it reduces the chance for tourists to learn about Jamaicans in general and traders in particular. In addition, the sector's encouragement of self-indulgent indolence increases the chance that tourists will behave in the hedonistic ways that dismay many Jamaicans. Taken separately, none of these policies and practices has a profound effect. However, taken together, they make it more likely that contact between tourists and traders will be fleeting and tense, that tourists will feel threatened and that traders will press every advantage that they can in the little time available to them.

We have described the stylised representations of two of the sets of people who concern us, tourists and traders. As we have shown, those in tourism present them as constraining the sector in its effort to be more beneficial for the people of Negril by reducing the isolation of tourists. In this rendering, those in tourism are stymied because the desires of tourists are incompatible with the avarice of traders. As we noted, however, when those in the sector treat tourist desires and trader behaviour as external constraints, they fetishise these sets of people. That is because their rendering and invocation ignore the context in which tourists and traders exist, and so ignore the ways that tourism practices shape tourists' desires and traders' behaviours, and do so in ways that help make the sector's arguments self-fulfilling, as well as self-serving.

Tourism, Tourists and Fishers

A different set of people that is portrayed as constraining the industry is Negril fishers. The sector's construction of fishers has two aspects that we will consider separately. The first revolves around including fishers and their beach in the activities of the sector, and so is concerned with the criticism that the sector excludes local people from the benefits of tourism. The second revolves around the ways that fishers use and affect the coastal waters, and so is concerned with the criticism that the sector harms the environment.

One of the ways that local people can be incorporated in tourism is by having places or practices that attract tourists. As we have described, there is little in Negril that has been around long enough to be counted as the sort of 'backstage' or 'authentic' Jamaica (Goffman, 1959; MacCannell, 1989 [1976]) that would be attractive. While it is not obviously old, the Negril fishing beach is an exception (see Figure 9.3). There, tourists can see people landing their catch and selling it, and see them repairing nets and traps. Tourists could eat at either of the two shops that sell local food, or just relax with a glass of rum and watch the world go by.[4] However, there are no tourists here: the fishing beach does not appear in tourist brochures or websites; it is effectively invisible, hidden behind the town's main craft market.

This absence of fishers from tourism is doubly problematic for the sector. They are not just local people who, critics say, ought not to be excluded. In addition, fishers of the sort based at the Negril beach appear to be especially significant throughout the English-speaking Caribbean (see Price, 1966). They are linked with important cultural values associated with freedom from slavery and oppression, and loom much larger in Caribbean people's minds and their understandings of the region and its history than their numbers would suggest. As part of this, they are routinely seen as deserving support.

Reflecting their expressed desire to spread the benefits of tourism, people in the sector were happy to talk about fishers. They said that there

Figure 9.3 The Negril main fishing beach

was real potential in the fishing beach, but did so in a way that indicated that they were constrained from realising that potential by fishers themselves. The immediate problem they described is the state of the fishing beach. It is no tourist attraction. Not only is it 'dirty' and filled with 'garbage', it is disorderly: 'unstructured', 'helter-skelter' and with 'no uniformity'. According to those in the sector, the fishing beach 'needs a lot of face lift' in order to make it look 'aesthetically appealing and more neat'. The 'shacks need to be a little bit more presentable' and 'not look like they will fall down over the tourists' heads'. Such opinions are not limited to Jamaica. For instance, Pugh (2005: 314) reports that those in tourism in Soufriere, in St Lucia, say that while the fishing beach there has potential, it is 'in need of help', and that those who use it need 'progress, structure and order'. Tidying the Negril fishing beach would improve things, but those in the sector said that if it were really going to attract tourists it ought to have a 'nice information booth', a 'rest room', a 'sitting area' and a 'cooler with Red Stripe', the ubiquitous Jamaican beer. Also, it should have 'bilingual crafts and fish vendors', fluent in German or Italian as well as English (92% of tourists to Jamaica come from English-speaking countries; 3% come from Germany and Italy combined; Bakker & Phillip, 2005: Fig. 11).

These changes are concerned with the appearance of and facilities at the fishing beach. For those in the sector, improving these would be important but inadequate, because that would not get at the real issue. The disorderly state of the beach reflects the disorderly state of the fishers, which also needs correction. People said that fishers were 'uneducated' and that the beach is home to illegal drugs, gambling and prostitution: 'even the police have a problem controlling things'. This undesirable state is made worse by the self-indulgent, indolent nature of the tourists that those in the sector say that they are obliged to serve. Those in tourism say that they 'can't ensure that guests feel free and relaxed' there. There would need to be a lot of changes, so that tourists would not experience 'a lot of harassment' at the beach.

In expressing concern about the disorderly nature of fishers, people in the sector were reflecting a distinction that is reported widely in the Caribbean. The classic name for it is 'reputation and respectability', which comes from Wilson's (1973) initial ethnographic description (various aspects of it are described in Abrahams, 1983; Austin, 1983, 1984; Besson, 1993; Browne, 2004; Miller, 1994; Olwig, 1995; Thomas, 2004; Wardle, 1999). The realm of reputation is one of playful disorder, spontaneity and equality, shading over into disruption and illegality. The realm of respectability is one of public morality, propriety, good order and respect for authority, shading over into authoritarianism. In identifying fishers and their beach as disorderly, disruptive and criminal, people in the sector place fishers in the realm of reputation, which is looked down on in public life. Implicitly, they also place themselves in the realm of respectability, which is esteemed in public life.

Not being respectable, fishers could not be expected to want to improve their fishing beach. Consequently, control of the beach needs to pass to the tourism sector, which would bring the sort of 'progress, structure and order' that Pugh's Soufriere informants said the beach there needed. As one person said, 'it takes management and organisation to make this an attraction', and management and organisation are two things that fishers cannot provide, but the tourism sector can. So, it would be necessary to employ resort guards who are 'hand selected and trained' to be 'watchful over the tourists'. With this sort of control in place, 'even locals would go there': as a man in the sector concluded, it would be a 'nice place to go on Sundays'.

As was the case with traders, here too those in the sector present themselves as happy to do more to spread the benefits of tourism among the people of Negril. However, once more they are constrained by sets of people who are portrayed as being, in crucial ways, independent of them and beyond their control: the desires of tourists on the one hand, and on the other hand, the nature of fishers, especially their disorganised inability to make themselves and their beach safe and respectable, much less appealing.

We said that fishers have a special place in people's self-conception and cultural past. Because of this, public criticism of them would be difficult, which helps explain the way that those in tourism made their criticisms. Most often, these were presented as coming from tourists or as being expressions of their interests. Personal critical comments, however, were made almost as slips of the tongue, followed by statements of regret and pleas not to be identified as the source of the criticism. Invoking tourists as the origin of these concerns allows those in tourism to avoid much of the opprobrium that overt criticism of fishers would generate.

This sort of use of tourists also appeared when those in tourism criticised fishers for their treatment of the coastal waters. 'The fishermen damage the environment by... not protecting the reef as they should'. A watersports manager said that their fishing boats (typically 25-foot long, narrow fibreglass boats with 40 hp outboard motors) are always 'leaking oil and gas into the water'. Reflecting the sector's professed commitment to environmental responsibility, and contrasting that with the supposed irresponsibility of fishers, a colleague continued, 'There is nothing more for the big guys [i.e. the tourism sector] to do. Now it is all up to the small guys [i.e. the fishers]'. Here, then, criticism of fishers highlights what is presented as the much better behaviour of those in the sector, and so helps protect tourism from criticism.

In interviews with people in the sector and in meetings of the Chamber of Commerce and of the Jamaica Hotel and Tourist Association, fishers were presented as harming the coastal waters and were criticised heavily for it. People said that fishers are indiscriminate, catching fish that are 'small', 'young', 'breeding', 'endangered' and 'pretty': they will 'kill anything to make some money'. They said that fishers violate the regulations of the Negril Marine Park by fishing in prohibited zones, that they use illegal fishing methods, that their techniques damage the coral. In all of this, people said, fishers care only for themselves and are heedless of the long-term consequences for the environment and the country, reliant on beach-front tourism. Perhaps reflecting the special cultural meaning of fishers, this criticism is commonly mixed with compassion. People routinely said that fishers have to make a living and they were fishing in Negril long before tourism developed. However, this compassion is limited. The fact that Negril started as a fishing site 'doesn't necessarily mean that it has to stay like that', and did not excuse what was seen as fishers' immorality and environmental damage.

There is a more dramatic example of this use of tourists as a vehicle for criticising fishers. In the late 1990s, there was a surreptitious war on fishers, when people cut the mesh on fish traps at sea. Several people in the sector mentioned this vandalism, especially those in the dive business. Overwhelmingly, they said that tourist divers were upset at seeing trapped fish and cut the mesh to release them. This may be. However, one knowledgeable person in the dive business said that the common story does not tell the truth. He said that it was people in the business who damaged the traps, because they were upset with what he said was the uncooperative attitude of fishers toward the Negril Marine Park. According to this man, the war ended when fishers finally retaliated, by cutting the lines anchoring the mooring buoys that the park had laid out in its waters and on which many of those in tourism relied. At that point, he said, those in the dive business decided that the conflict had to stop.

Once more, this construction of fishers treats them as autonomous by ignoring the ways that the sector is implicated in shaping them. This is most obvious when fishers' violation of marine park regulations is presented to indicate their inherent unruliness and immorality. Such a presentation ignores how the park was developed (aspects are described in Carrier, 2003). Initial agitation for the park was from those in the sector; the park's policies generally seek to exclude fishers while welcoming tourists; the park is much more likely to criticise and seek to regulate fishers than tourism. Given this, fishers would be unlikely to view the park favourably or see its policies and regulations as straightforward reflections of environmental necessity. Here, then, the sector's relationship with the park, combined with the park's dealings with fishers, helps to generate what those in the sector portray as signs of immorality and unruliness among an autonomous set of people.

Other aspects of the construction of fishers identify certain activities as characteristic of them and hence grounds for criticism. These renderings do not mislead by fetishising fishers, but by selectively comparing fishers and the tourism sector. We have already mentioned that some people pointed to fishers' outboard motors and the pollution that they generate as evidence that fishers are irresponsible. This ignores the sector's equally polluting jet skis. Similarly, the claim that fishers damage reefs with their traps ignores the fact that hotel guests damage reefs when snorkelling (for a further parallel with Soufriere, see Pugh, 2005: 316). Condemnation of the immorality of fishers ignores the immorality of hotel guests. These renderings shape the image of the fisher in an implicit way, by their silences. Behaviour that merits criticism when found among fishers is passed over in silence when found in tourism.

Conclusion

We have addressed the ways that those in tourism construe and represent tourists, traders and fishers, and do so in ways that protect the sector from those who criticise it for its social and environmental effects. Our approach to these constructions echoes Marx's (1867) discussion of fetishism. As Marx argued that the understanding of commodities ignored the social relations and practices that produced them, so the construction of tourists, traders and fishers ignores the social relations and practices that shape them. As Marx argued that commodities were thereby fetishised, seen as confronting people as distinct and autonomous entities, so tourists, traders and fishers are fetishised, presented as distinct and autonomous sets of people who confront and constrain the tourism sector in ways that the critics do not recognise. The result is a construction that helps secure the sector's position in Jamaica's political economy.

We want to close by briefly considering two additional aspects of the construction of these groups. They are pertinent because they further illustrate the issue we laid out at the start of this chapter, the ways that people's statements about the nature of the world, which we take as reflections of their knowledge of it, can secure their interests and authority in relation to others.

The first aspect concerns the issue of the state of the natural environment and the environmental credentials of the sector. We have already described the way that people in the sector protect themselves from criticism: they focus attention on fishers' activities in the coastal waters, which they say are more harmful than the activities of the sector and tourists. Here, however, we want to point to a more thorough-going definition of the natural surroundings that is implicit in what people say about fishers. For those in the sector, the coastal waters need to be maintained in ways that will continue to attract tourists. If they are maintained that way, the waters will benefit Jamaica, reliant as it is on tourism. A responsible attitude towards the surroundings, then, is one that seeks to assure its attractiveness. Opposed to this is the attitude of those stereotyped, imagined fishers, whose heedless use of the waters to secure their subsistence reduces the waters' attractiveness and thus harms the country as a whole. Such a construction conflates the environmental condition of the coastal waters with their attractiveness to tourists. The result is that the environmental well-being of the waters is defined in terms of the interest of the sector (see Carrier, 2003: 216–222), itself defined as the interest of Jamaica.

The second aspect that we describe is political-economic rather than environmental. We noted that people in tourism construct traders as people who stand outside of tourism and hinder its ability to benefit the country. However, traders are engaged in commercial transactions with tourists and are, in fact, part of the country's tourism economy; in other words, part of the tourism sector rather than outside of it. By placing such people outside the pale, the construction of the trader implicitly defines the tourism sector in a way that restricts it to fairly large hotel corporations and the companies with which they do business, marginalising and even demonising the rest, rather as Malcolm Crick (1994: 159–160) describes for Sri Lanka. However, critics see things differently, saying that large corporations prevent the economic benefits of tourism spreading to ordinary people. In an effort to increase that spread, some have advocated community tourism (Duperly-Pinks, 2002; Hayle, 2000), which defines the sector more broadly. Whether or not it can produce the results claimed for it, its advocates urge it precisely because it benefits small businesses: guest houses, small restaurants and traders. This advocacy has been effective to the extent that expanding community tourism is included among the goals of the tourism master plan (Jamaica, 2003).

The stakes involved are high. The government is determined to expand tourism, so those who are within the sector, as it is conventionally defined, can look forward to support denied to those outside of it, support that would be diluted if the conventional definition were expanded. Given these stakes, it would not be surprising if efforts to extend community tourism resulted in little more than established companies being awarded contracts to take over things like the Negril fishing beach, to make them 'aesthetically appealing and more neat', a place where 'guests will feel free and relaxed'. But, whatever the outcome, government support for community tourism shows the importance of tourism's critics, and the importance for the sector of the ways that it presents itself, tourists, traders and fishers.

Acknowledgements

We want to thank David Dodman for his generous help with information on the early history of Negril. This research was funded by the Economic and Social Research Council of the UK, and was part of the project 'Conflict in environmental conservation: A Jamaican study' (RES 000-23-0396). Although Sommer and Carrier wrote this paper, inevitably it relies on the research of, and our conversations with, the other two researchers on this project, Andrew Garner and Monica

Lorenzo Pugholm. Also, it reflects the helpful comments on earlier versions made by David Dodman, Andrew Garner, Deborah Gewertz, Josiah Heyman, Mark Johnson, Donald Macleod, Mimi Sheller and Paige West. We are grateful for their time and thoughts.

Notes

1. Other scholars have pointed to aspects of the processes and issues that concern us (e.g. Boissevain, 1996; Stonich, 1998; D. Wilson, 1994), though Stronza's (2001) review of the anthropology of tourism indicates that there is little such work. Of course, since then her review *Contesting the Foreshore* (Boissevain & Selwyn, 2004) has appeared. It illustrates the sorts of issues that concern us, albeit using a somewhat different theoretical perspective and describing primarily the Mediterranean.
2. The 'chosen few' who benefit from tourism tend to be white, at least in Jamaican terms, which are complex (see Austin-Broos, 1994). However, the people we studied did not invoke race. It may have influenced people's perceptions of each other, but it was not part of what they said about each other. Wealth, education and nationality were what concerned them.
3. Unless otherwise indicated, quoted words and phrases are from interviews with tourists and those in the tourism sector.
4. It is not clear that fishers want tourist visitors. Common opinion among them was that tourists are frequently intrusive to the point of being rude, and that their public semi-nudity is immoral. The actual opinions of fishers are insignificant in terms of the sector's stylised rendering of them: on this and other matters, fishers' views were assumed by those in tourism, not investigated.

References

Abrahams, R.D. (1983) *The Man of Words in the West Indies: Performance and the Emergence of Creole Culture.* Baltimore, MD: Johns Hopkins University Press.
Aiken, K.A. and Haughton, M.H. (1987) Regulating fishing effort: The Jamaican experience. *Proceedings of the Annual Gulf and Caribbean Institute* 40, 139–150.
Austin, D.J. (1983) Culture and ideology in the English-speaking Caribbean: A view from Jamaica. *American Ethnologist* 10, 223–240.
Austin, D.J. (1984) *Urban Life in Kingston, Jamaica: The Culture and Class Ideology of Two Neighborhoods.* New York: Gordon and Breach.
Austin-Broos, D.J. (1994) Race/class: Jamaica's discourse of heritable identity. *Nieuwe West-Indische Gids* 68, 213–233.
Bakker, M. and Phillip, S. (2005) *Travel and Tourism – Jamaica – February 2005.* London: Mintel International Group Ltd. (n.p.)
Bartilow, H.A. (1997) *The Debt Dilemma: IMF Negotiations in Jamaica, Grenada and Guyana.* London: Macmillan.
Bélisle, F.J. (1983) Tourism and food production in the Caribbean. *Annals of Tourism Research* 10, 497–513.
Besson, J. (1993) Reputation and respectability reconsidered: A new perspective on Afro-Caribbean peasant women. In J. Momsen (ed.) *Women and Change in the Caribbean* (pp. 15–37). London: James Currey.

Black, S. (2001) *Life and Debt.* Kingston: Tuff Gong Pictures.

Blackstone Corporation (2001) *Sustainable Tourism Project Proposal Preparation, Portland. Phase I Briefing Document: Overview of Issues, Funding Opportunities and Workshop Suggestions.* Toronto: Blackstone Corporation.

Bloom, D.E., Mahal, A., King, D., Henry-Lee, A. and Castillo P. (2001) *Occasional Paper: Globalization, Liberalization and Sustainable Human Development: Progress and Challenges in Jamaica.* Geneva: United Nations Conference on Trade and Development, and United Nations Development Programme. On WWW at http://www.unctad.org/en/docs/poedmm176.en.pdf. Accessed 15.5.06.

Boissevain, J. (ed.) (1996) *Coping with Tourists: European Reactions to Mass Tourism.* Oxford: Berghahn.

Boissevain, J. and Selwyn, T. (eds) (2004) *Contesting the Foreshore: Tourism, Society, and Politics on the Coast.* (MARE publication series 2.) Amsterdam: Amsterdam University Press.

Bolles, A.L. (1992) Sand, sea, and the forbidden. *Transforming Anthropology* 3 (1), 30–34.

Bourdieu, P. (1977) *Outline of a Theory of Practice.* Cambridge: Cambridge University Press.

Brohman, J. (1996) New directions in tourism for Third World development. *Annals of Tourism Research* 23, 48–70.

Browne, K.E. (2004) *Creole Economics: Caribbean Cunning under the French Flag.* Austin, TX: University of Texas Press.

Burke, L., Kura, Y., Kassem, K., Revenga, C., Spalding, M. and McAllister, D. (2000) *Research Report: Pilot Analysis of Global Ecosystems: Coastal Ecosystems.* Washington, DC: World Resources Institute.

Burke, R.I. (2005) *Environment and Tourism: Examining the Relationship between Tourism and the Environment in Barbados and St. Lucia.* (Sustainability impact assessment of the new Economic Partnership Agreements between the ACP & the GCC States and the EU.) Neuilly-sur-Seine: PriceWaterhouseCoopers.

Caribbean Association for Feminist Research and Action (2002) Jamaican women cry for action not 'a bag a mouth'. On WWW at http://www.cafra.org/article.php3?id_article = 79. Accessed 11.5.06.

Carrier, J.G. (2003) Biography, ecology, political economy: Seascape and conflict in Jamaica. In A. Strathern and P.J. Stewart (eds) *Landscape, Memory and History* (pp. 210–228). London: Pluto Press.

Chambers, D. and Airy, D. (2001) Tourism policy in Jamaica: A tale of two governments. *Current Issues in Tourism* 4 (2–4), 94–120.

Cheong, S-M. and Miller, M.L. (2000) Power and tourism: A Foucauldian observation. *Annals of Tourism Research* 27, 371–390.

CIDA (2002) *INC – Gender Profile: Jamaica.* Gatineau, Quebec: Canadian International Development Agency. On WWW at http://www.cida.gc.ca/.../3f70b80bd2c2e2ff8525664200403dad/6718d10a2e97648085256bf90047ee94?OpenDocument. Accessed 27.4.06.

Crick, A.P. (2002) 'Smile, you're a tourism employee!' Managing emotional displays in Jamaican tourism. In I. Boxill, O. Taylor and J. Maerk (eds) *Tourism and Change in the Caribbean and Latin America* (pp. 162–178). Kingston: Arawak Publications.

Crick, M. (1989) Representations of international tourism in the social sciences: Sun, sex, sights, savings, and servility. *Annual Review of Anthropology* 18, 307–344.

Crick, M. (1994) *Resplendent Sites, Discordant Voice: Sri Lankans and International Tourism.* Chur: Harwood Academic.

Dann, G. (1996) The people of tourist brochures. In T. Selwyn (ed.) *The Tourist Image: Myths and Myth Making in Tourism* (pp. 61–81). Chichester: John Wiley and Sons.

Davies, T. and Cahill, S. (2000) *Environmental Implications of the Tourism Industry.* (Discussion paper 00-14.) Washington, DC: Resources for the Future. On WWW at http://ideas.repec.org/p/rff/dpaper/dp-00-14.html. Accessed 15.5.06.

Duperly-Pinks, D. (2002) Community tourism: 'Style and fashion' or facilitating empowerment? In I. Boxill, O. Taylor and J. Maerk (eds) *Tourism and Change in the Caribbean and Latin America* (pp. 137–161). Kingston: Arawak Publications.

Echtner, C.E. and Prasad, P. (2003) The context of Third World tourism marketing. *Annals of Tourism Research* 30, 660–682.

European Commission (1999–2000) Tourism: Jamaica is here to stay. (Country report: Jamaica.) *The ACP-EU Courier 178* (Dec-Jan), np. On WWW at http://ec.europa.eu/comm/development/body/publications/courier/index_178_en.htm. Accessed 15.5.06.

Eversley, M. (2003) Some Jamaicans feel the island's majority of all-inclusive resorts shuts out local business. *The Atlanta Journal-Constitution* (22 September). On WWW at http://www.hotel-online.com/News/PR2003_3rd/Sep03_AllInclusive.htm. Accessed 15.5.06.

Gmelch, G. (2003) Island tourism. In G. Gmelch (ed.) *Behind the Smile: The Working Lives of Caribbean Tourism* (pp. 1–24). Bloomington, IN: Indiana University Press.

Goffmann, E. (1959) *The Presentation of Self in Everyday Life.* New York: Doubleday.

Grandoit, J. (2005) Tourism as a development tool in the Caribbean and the environmental by-products: The stresses on small island resources and viable remedies. *Journal of Development and Social Transformation* 2, 89–97.

Haley, M. and Clayton, A. (2003) The role of NGOs in environmental policy failures in a developing country: The mismanagement of Jamaica's coral reefs. *Environmental Values* 12, 29–54.

Hayle, C. (2000) Community tourism in Jamaica. In J. Maerk and I. Boxill (eds) *Turismo en el Caribe – Tourism in the Caribbean* (pp. 165–176). México, DF: Plaza y Valdés Editores.

Island Resources Foundation (1996) *Tourism and Coastal Resources Degradation in the Wider Caribbean.* St Thomas, VI: Island Resources Foundation. On WWW at http://www.irf.org/irtourdg.html. Accessed 15.5.06.

Jamaica (2003) *Sustainable Tourism Master Plan.* Kingston: Ministry of Tourism and Sport. On WWW at http://www.jsdnp.org.jm/Tourism%20Master%20Plan.PDF. Accessed 11.5.06.

Jamaica Environment Trust and Northern Jamaica Conservation Association (2005) *Review of the Environmental Impact Assessment Done by Environmental Solutions Limited of the Proposed Bahia Principe Hotel Resort Development Pear Tree*

Bottom, St. Ann Jamaica. Kingston: Jamaica Environment Trust. On WWW at http://www.jamaicancaves.org/bahia-principe-runaway-bay.htm. Accessed 15.5.06.

Kozyr, E. (2000) The negative effects of tourism on the ecology of Jamaica. On WWW at http://www.saxakali.com/caribbean/ekozyr.htm. Accessed 15.5.06.

MacCannell, D. (1989 [1976]) *The Tourist: A New Theory of the Leisure Class* (rev. edn). New York: Schocken.

McLaren, D. (1998) *Rethinking Tourism and Ecotravel: The Paving of Paradise and What You Can Do To Stop It*. West Hartford, CN: Kumarian Press.

von Maffei, P. (2000[?]) Will Jamaica self-destruct? On WWW at http://debate.uvm.edu/dreadlibrary/vonmaffei.html. Accessed 15.5.06.

Marx, K. (1867) The fetishism of commodities and the secret thereof. *Capital, Volume 1* (Part 1, Chap. 1, Sect. 4). Hamburg: Meissner

Miller, D. (1994) *Modernity: An Ethnographic Approach*. Oxford: Berg.

Olsen, B. (1997) Environmentally sustainable development and tourism: Lessons from Negril, Jamaica. *Human Organization* 56, 285–293.

Olwig, K.F. (1995) Cultural complexity after freedom: Nevis and beyond. In K.F. Olwig (ed.) *Small Islands, Large Questions: Society, Culture and Resistance in the Post-Emancipation Caribbean* (pp. 100–120). London: Frank Cass.

O'Neil, M. (2003) *Tourism Maturity and Demand: Jamaica*. Kingston: Bank of Jamaica.

Panos Institute (1998) *Improving Training and Public Awareness on Caribbean Coastal Tourism*. Kingston[?]: UNEP[?]. On WWW at http://www.cep.unep.org/issues/panos.pdf. Accessed 15.5.06.

Pattullo, P. (1996) *Last Resorts: The Cost of Tourism in the Caribbean*. Kingston: Ian Randle.

Price, R. (1966) Caribbean fishing and fishermen: A historical sketch. *American Anthropologist* 68, 1364–1383.

Pruitt, D. and LaFont, S. (1995) For love and money: Romance tourism in Jamaica. *Annals of Tourism Research* 22, 422–439.

PSOJ (1997) *Annual Report of the Executive Committee: Selected Topics – Tourism*. Kingston: Private Sector Organization of Jamaica. On WWW at http://www.psoj.org/reports/annual97Tourism.pdf. Accessed 15.5.06.

PSOJ (1999) *Annual Report of the Executive Committee: Selected Topics – Tourism*. Kingston: Private Sector Organization of Jamaica. On WWW at http://www.psoj.org/reports/annual99Tourism.pdf. Accessed 15.5.06.

Pugh, J. (2005) The disciplinary effects of communicative planning in Soufriere, St Lucia: Governmentality, hegemony and space-time-politics. *Transactions of the Institute of British Geographers* (N.S.) 30, 307–321.

Sheller, M. (2004) Demobilizing and remobilizing Caribbean paradise. In M. Sheller and J. Urry (eds) *Tourism Mobilities: Places to Play, Places in Play* (pp. 13–21). London: Routledge.

Sheller, M. and Urry, J. (2004) Introduction: Places to play, places in play. In M. Sheller and J. Urry (eds) *Tourism Mobilities: Places to Play, Places in Play* (pp. 1–12). London: Routledge.

Stonich, S.C. (1998) Political ecology of tourism. *Annals of Tourism Research* 25, 25–54.

Strathern, M. (1996) Cutting the network. *Journal of the Royal Anthropological Institute* (N.S.) 2, 517–535.

Stronza, A. (2001) Anthropology of tourism: Forging new ground for ecotourism and other alternatives. *Annual. Review of Anthropology* 30, 261–283.

Taylor, F. (1993) *To Hell with Paradise: A History of the Jamaican Tourist Industry.* Pittsburgh, PA: University of Pittsburgh Press.

Taylor, O.W. (2002) Worker protection and termination of the contract of employment in the Caricom tourism sector. In I. Boxill, O.W. Taylor and J. Maerk (eds) *Tourism and Change in the Caribbean and Latin America* (pp. 192–206). Kingston: Arawak Publications.

Thomas, D.A. (2004) *Modern Blackness: Nationalism, Globalization, and the Politics of Culture in Jamaica.* Durham, NC: Duke University Press.

Thomas-Hope, E. and Jardine-Comrie, A. (2005) Valuation of environmental resources for tourism: The case of Jamaica. IRFD World Forum on Small Island Developing States. Cambridge, MN: International Research Foundation for Development. On WWW at http://irfd.org/events/wfsids/virtual/papers/sids_ethomashope.pdf. Accessed 15.5.06.

UNEP (2002) Negative socio-cultural impacts from tourism. United Nations Environment Programme. On WWW at http://www.uneptie.org/pc/tourism/sust-tourism/soc-drawbacks.htm. Accessed 11.5.06.

Wang, N. (2000) *Tourism and Modernity: A Social Analysis.* Amsterdam: Pergamon.

Wardle, H. (2000) *An Ethnography of Cosmopolitanism in Kingston, Jamaica.* Lewiston, NY: Edwin Mellen Press.

Wilson, D. (1994) Probably as close as you can get to paradise: Tourism and the changing image of Seychelles. In A.V. Seaton (ed.) *Tourism: The State of the Art* (pp. 765–774). Chichester: Wiley.

Wilson, P.J. (1973) *Crab Antics: The Social Anthropology of English-Speaking Negro Societies of the Caribbean.* New Haven, CT: Yale University Press.

Witter, M. (2005) *Trade Liberalization: The Jamaican Experience.* Geneva: United Nations Conference on Trade and Development, Trade Analysis Branch. On WWW at http://192.91.247.38/tab/events/namastudy/fullreport-version14 nov-p202-231.pdf. Accessed 15.5.06.

Epilogue

Chapter 10
Power in Tourism: Tourism in Power

C. MICHAEL HALL

The issue of power is one that has become increasingly significant in the study of tourism and, from a low base, has resulted in several recent books and papers, including the current volume (see also Church & Coles, 2007). However, even though there is now belated recognition of power as an issue in tourism studies, it remains a relatively peripheral concern in most research. In one sense, this is somewhat surprising given that the regulation of movement is one of the most basic elements of tourism and tourist behaviour and that no right exists to enter into another country. Yet, in another sense, it may reflect a wider apolitical and often uncritical discourse that dominates much of tourist studies, which is grounded in an inherent managerialism and economism that understands the study of tourism as preparing students for employment in the tourism industry and tourism research as being 'for' such industry.

Interest in power may also reflect a maturing of tourism as a disciplinary field within the social sciences (Hall, 1994). Although more likely it is a result of the wave of interest in Foucault in tourism studies from the mid-1990s on, initiated in part by the work of Urry (1990)[1] and, if one were to be a touch cynical, in both cases often frequently cited, occasionally read and little understood. Perhaps describing the work of Foucault as 'little understood' may be slightly unfair (though see Lukes (2005) and below for similar assertions). Instead, it may be more accurate to describe the situation as one in which Foucault's work on power has often been utilised in tourism without a broader understanding of how his work fits in with wider debates on power, especially with respect to areas such as tourism development and communities. Instead, this chapter seeks to take a different tack and rather than reifying Foucault, it seeks to connect the study of tourism to the range of approaches that can and have been used to understand power relations. This is regarded as being significant for the inter-relationships between tourism, culture and power, because the exercise of power is witnessed not just in the production and reproduction of elements of culture, but also in the very direct exercise of power by individuals and in the shaping of institutions and the rules of the game.

The Study of Power

The dimensions of power and its exercise has been a subject of discussion in the social sciences since the days of Ancient Greece. Russell (1938: 10) stated in his study of power: 'The fundamental concept in social science is Power in the sense in which Energy is the fundamental concept in physics'. Like energy, power has many forms, such as wealth, armaments and influence on opinion. No one of these can be regarded as subordinate to any other, and there is no form from which the others are derivable. Concerns over the relative rights of the state and of individuals and the relationship of the individual to the state have been a cornerstone of political philosophy little touched on in tourism. Even the structural analyses of Marx and Gramsci, which were so significant for much of the social sciences for many years, have only been lightly touched on in tourism (e.g. MacCannell, 1976, 1999; Watson & Kopachevsky, 1994; Ateljevic & Doorne, 2002) and have arguably been more important historically in cognate fields such as leisure (e.g. see Rojek, 1985, 1995). Perhaps more significantly, and even surprisingly given the focus on community in much tourism planning studies of the 1980s and 1990s, is the almost total absence of appreciation of the power structure debates surrounding power and elites in American political science from the 1950s on and later connected to British debate. Although perhaps lacking the exoticism or density of translated French political philosophy, the issues of structure, the powerless and domination that lay at the heart of the Anglo-American debate nevertheless remain important for understanding the concept of power and its relevance to tourism and culture.

Mills' (1956: 3) influential book on *The Power Elite* argued: 'The powers of men are circumscribed by the everyday worlds in which they live, yet even in these rounds of job, family and neighbourhood they often seem driven by forces they can neither understand nor govern'. According to Mills, the 'power elite',

> are in positions to make decisions having major consequences. Whether they do or do not make such decisions is less important than the fact that they do occupy such pivotal positions; their failure to act, their failure to make decisions, is itself an act that is often of greater consequence than the decisions they do make. For they are in command of the major hierarchies and organizations of modern

society. They run the big corporations. They run the machinery of state and claim its prerogatives. They direct the military establishment. They occupy the strategic command posts of the social structure, in which are now centred the effective means of the power and the wealth and the celebrity they enjoy. (Mills, 1956: 3–4)

Similarly, Hunter (1953), writing on decision making in community power structures, argued there was substantial agreement among community leaders most of the time on the big issues of community and institutional culture. Noting how community organisations

are controlled by men who use their influence in devious ways, which may be lumped under the phrase "being practical", to keep down public discussion on all issues except those that have the stamp of approval of the power group. (Hunter, 1953: 249)

The conclusions of Hunter and Mills that elites dominated national and local governments in a very direct manner caused a major reaction in the USA, where power is often perceived as being pluralistic in fashion because of a range of visible interest groups being visible in the policy-making process, as well as the right to assemble, freedom of speech, elections, competitive political parties and the existence of a market economy (Domhoff, 2007). The most significant academic reaction came from Dahl (1961), who argued that power was intentional and active and related to several, separate, single issues and bound to the local context of its exercise (Lukes, 2005). The emphasis by writers such as Dahl on overt preferences of interest groups and the exercise of political power served to strengthen the pluralistic conception of power that, 'since different actors and different interest groups prevail in different issue-areas, there is no overall "ruling elite"' (Lukes, 2005: 5). Such an overt interest group model of power structures was influential within tourism, especially in the work of Murphy (1985), and such a 'participatory' model of how destination communities operated dominated in tourism planning for much of the 1980s and 1990s with only little criticism of the underlying assumptions with respect to how power may be used to exclude other interests from decision making. This is also not to say that the study of overt power and interest is without value for understanding tourism and cultural relationships. For example, the increased legal standing of indigenous peoples and their culture in countries such as Australia is clearly related to the exercise of power and the capacity to have some issues kept at a high level on the political agenda. As an example, on 3 June 1992, the Australian High Court ruled that the land title of Australia's indigenous peoples, the Aborigines and Torres Strait Islanders, is recognised at common law and that indigenous land title, or native title, stems from the continuation within common law of their rights over lands and inshore waters that pre-date European colonisation of Australia. Although the case did not apply to private lands, it did apply to crown lands, national parks and leased crown lands that recognised Aboriginal use. Control of land is clearly an important factor in tourism development (Hall, 2007b). As Liam Myer, executive officer of the Djabiluka Aboriginal Association in the Kakadu region stated with respect to the establishment of joint ventures between Aboriginal and non-Aboriginal groups: 'The power that the Aboriginal traditional owners have is access to the land, access to the site' (quoted in Langton & Palmer, 2003: np). Nevertheless, an obvious follow-up question of power that arises from such a decision is why did it take so long?

The power structure debate is grounded in broader questions as to how power is conceptualised and how it can be studied. Key questions with respect to power structure research include (1) what organisation, group or class in the social structure under study receives the most of what people seek and value (*who benefits*)? (2) what organisation, group or class is over-represented in key decision-making positions (*who sits*)? (3) what organisation, group or class wins in the decisional arena (*who wins*)? and (4) who is thought to be powerful by knowledgeable observers and peers (*who has a reputation for power*)? (Domhoff, 2007). Questions that reflect Lasswell's (1936) comment about politics: politics is about power, who gets what, where, how and why. Unfortunately, in tourism studies such questions are often never asked (Hall 1994, 2007a).

Conceptualising power, however, presents substantial difficulties. Indeed, it has been characterised as an 'essentially contested concept' (Gallie 1955–1956).

> Power is often spoken of as if it were a unitary and independent force, sometime incarnated in the image of a giant monster such as Leviathan or Behemoth, or else as a machine that grows in capacity and ferocity by accumulating and generating more powers, more entities like itself. Yet it is best understood neither as an anthropomorphic force nor a giant machine but as an aspect of all relations among people. (Wolf, 1999: 4)

Considering power in relational terms highlights the manner in which power works differently in interpersonal and institutional relations, as well as society as a whole. Wolf (1999) distinguished between four different modalities with respect to the relationship between power and social relations. First is the capability of an individual in a Nietzschean sense; second is power manifested in interactions between people as refers to the Weberian notion of the ability of an ego to impose its will to social action upon an alter; third is the power, referred to by Wolf as tactical or organisational power, which controls the contexts or instrumentalities by which individuals or groups direct or circumscribe the actions of others; fourth is the role of structural power, which is the power manifest in relationships that not only operates within settings and domains, but also organises the settings themselves. This last mode of power is akin to what Foucault (1984) described as 'governance', the exercise of 'action upon action' in which he sought to determine those relations that structured consciousness, an area also significant to Marxist analysis of the structural relations of power. Although Wolf's different modalities of power are not shared by all power theorists, it does share some significant common ground. In particular, it highlights the critique of the advocates of pluralism that such a conception of power was 'too narrowly drawn' (Bachrach, 1967: 87). Bachrach and Baratz (1962) argued that power had a 'second face' that was not perceived or understood by pluralists nor detectable by their modes of inquiry. Instead,

> power was not solely reflected in concrete decisions; the researcher must also consider the chance that some person or association could limit decision-making to relatively non-controversial values, by influencing community values and political procedures and rituals, notwithstanding that there are in the community serious but latent power conflicts. (Lukes, 2005: 6)

The Second Face of Power

Bachrach and Baratz (1970) identified two major weaknesses in the pluralist approach to power: first, it did not provide for the fact that power may be exercised by confining the scope of political decision making; second, they argued that the pluralist model provided no criteria for determining the significant issues. Therefore, two-dimensional views of community power structures focus on decision making and non-decision making and observable (overt and covert) conflict (Bachrach & Baratz, 1962, 1970). The concept of non-decision making is invaluable in examining the pathways of tourism decision making (Hall & Jenkins, 1995), although little applied (i.e. Doorne, 1998). Bachrach and Baratz (1970: 44) defined a non-decision as 'a decision that results in suppression or thwarting of a latent or manifest challenge to the values or interests of the decision-maker'. A non-decision is, therefore, a means by which demands for change in the existing allocation of benefits and privileges in a community can be suffocated before they are even voiced; or kept covert, or killed-off before they gain access to the relevant decision-making arena; or, failing all these things, maimed or destroyed in the implementation stage of the policy process (Lukes, 1974; Hall, 2008). Non-decision making exists 'to the extent that a person or group – consciously or unconsciously – creates or reinforces barriers to the public airing of political conflicts, that person or group has power' (Bachrach & Baratz, 1970: 8). Non-decision making allows political actors, organisations and collectives to 'leave selected topics undiscussed for what they consider their own advantage' (Holmes, 1988: 22). Such a perspective can be important for gaining a better understanding of the actions of government and agencies with respect to the management of contentious heritage, for example by ensuring that a policy-making system favours one group (often development interests), while restricting the capacity of others to act.

The critique of pluralism is closely linked to Schattschneider's (1960) concept of the 'mobilisation of bias'. According to Schattschneider (1960: 71), 'all forms of political organization have a bias in favour of some kinds of conflict and the suppression of others because organization is the mobilization of bias. Some issues are organized into politics while others are organized out'. For example, to return to the Australian Aboriginal example noted above, it should be remembered that it was not until 1967 that Australians voted in a national referendum to change the national constitution in order to allow the federal government to make laws for Aborigines and to include them in the census, an amendment that provided Aborigines with full voting rights. Until the 1960s, therefore, Aboriginal issues (along with many other indigenous groups in developed countries) had been mobilised out of politics by denial of access to political structures and institutions (Hall, 2007b).

Bachrach and Baratz (1970: 11) similarly stressed the importance of an analysis of the 'mobilization of bias', which is 'the dominant values and the political myths, rituals, and institutional practices which tend to favour the vested interests of one or more groups, relative to others'. Non-decision making is regarded as the primary method for sustaining a given mobilisation of bias. The limiting of options in referenda, for example, is a classic example of non-decision making when electors are given a number of options with respect to development or other proposals. In the case of the 2002 British Columbia (BC) referendum on Native land claims, the wording of the referendum was described as 'amateurish' by the Angus Reid polling company, 'but the basic message was that it was unjust to put the rights of a minority group to the vote of a majority and that the questions being asked were designed to garner a "yes" vote' (Rossiter & Wood, 2005: 360). Indeed, a yes vote was the result, with voters responding 'yes' to the various questions, ranging from 84.52% (Question 1) to 94.50% (Question 4); over 20,000 votes were not considered as they did not meet the requirements of the Treaty Negotiations Referendum Regulation (Elections BC, 2002). Such 'spoiled' ballots were likely to be protest votes as a result of a campaign by the Union of BC Indian Chiefs. Elections BC also received letters and written comments that

> expressed concern that there was no mechanism to cast a "protest vote", or to have a means of influencing the outcome of the referendum other than to vote Yes or No... Similar concerns that there is not a "none of the above" option on election ballots have also been expressed by voters. (Elections BC, 2002: 7, 8)

However, just as importantly with respect to issues of mobilisation of bias and legitimacy, only 35.8% of registered voters actually returned ballots.

A variation of non-decision making is the concept of non-implementation in which, although policy is developed or regulation enacted, it is not actually enforced (Mokken & Stokman, 1976). For example, in New Zealand, successive governments since the early 1980s developed tourism policies that aimed to integrate Maori into tourism policy making as well as assist in the development of Maori tourism products. However, it was not until the creation of a coalition agreement between the National Party and New Zealand First in 1996 that the New Zealand Tourism Board had to formally consult with the Maori Ministry over Maori tourism issues (Hall, 2007b).

The practice of non-decision making and the mobilisation of bias well illustrates the rather romantic view that exists in some quarters with respect to the capacities of communities to undertake collaborative decision making. Instead, communities are not the embodiment of innocence;

> on the contrary, they are complex and self-serving entities, as much driven by grievances, prejudices, inequalities, and struggles for power as they are united by kinship, reciprocity, and interdependence. Decision-making at the local level can be extraordinarily vicious,

personal, and not always bound by legal constraints. (Millar & Aiken, 1995: 629)

Indeed, it is somewhat ironic that such cautionary words have not been heeded in tourism planning and policy making, given the interest by government, industry and many academics in greater collaborative decision making in tourism, with many such studies noting whose views have been included while failing to observe those who have been left out (Hall, 2008).

The focus on the mobilisation of bias and the behind-the-scenes agenda setting in the practice of non-decision making has led a number of scholars to investigate the value of Gramsci's concept of hegemony and the way in which the manufacture of culture constituted the 'mode of class rule secured by consent' (Anderson, 1976–1977: 42). Such voluntary consent could vary in intensity,

> On one extreme, it can flow from a profound sense of obligation, from wholesale internalisation of dominant values and definitions; on the other from their very partial assimilation, from an uneasy feeling that the status quo, while shamefully iniquitous, is nevertheless the only viable form of society. (Fernia, 1981: 39)

For example, to refer again to the situation in British Columbia, Rossiter and Wood (2005) argue that the Provincial Government sought to develop a set of indigenous policies that endeavoured to avoid overt protest, but which remained committed to private economic development with respect to First Nations. The Provincial Government had been promoting the benefits of the 2010 Winter Olympic Games for First Nations peoples, including a programme to boost 'aboriginal tourism'.

> As we invest in First Nations by creating new opportunities, one of our priorities is to ensure that we are matching skills training with areas of greatest need in our economy – clearly tourism is one of those. There are enormous openings emerging in aboriginal tourism as we prepare for the Olympics and we are working to support these. This new program will help build management and administrative skills for First Nations and enhance the entrepreneurial spirit that every successful industry needs. (Ministry of Advanced Education and Treaty Negotiations Office, 2004)

Yet, the Province also simultaneously denied 'the complexity that lies behind First Nations' assertions of land title and rights to self-government' and instead indicated a desire to attract investment 'within the logic of neo-liberalism: as is demonstrated by the "Aboriginal tourism" program' (Rossiter & Wood, 2005: 365).

Such notions of hegemony proved extremely influential in the development of a third face or dimension of power. The third dimension arises out of Lukes' (1974) critique of non-decision making with respect to non-intentional use of power as well as the role of power in shaping preferences. The three-dimensional view of power

> allows for consideration of the many ways in which potential issues are kept out of politics, whether through the operation of social forces and institutional practises or through individuals' decisions. (Lukes, 1974: 24)

Lukes (1974: 22) argued that Bachrach and Baratz (1970) did not recognise that the phenomenon of collective action is not necessarily 'attributable to particular individual decisions or behaviour, nor that the mobilisation of bias results from the form of *organisation*, due to "systemic" or organisational effects'. He then went on to emphasise the role that power has in shaping human preferences by arguing, 'to assume that the absence of grievances equals genuine consensus is simply to rule out the possibility of false or manipulated consensus by definitional fiat' (Lukes, 1974: 24): 'A may exercise power over B... by influencing, shaping or determining his very wants' (Lukes, 1974: 23). To Lukes (1974: 23), such an approach was 'the most effective and insidious use of power'.

Revisiting Lukes

Examination of the three-dimensional view of power may appear to be quite problematic. After all, 'how can one study, let alone explain, what does not happen?' (Lukes, 1974: 38). Nevertheless, as Crenson (1971: vii) recognised, the way 'things do not happen' is as important as what does: 'the proper object of investigation is not political activity but political inactivity'. Indeed, Lukes (1974) argued that third-dimensional power may be recognised when it is not in accordance with an individual or group's 'real interests' (Lukes, 1974: 24–25) (a concept similar to Marxian ideas of false consciousness (Hyland, 1995)). Nevertheless, Lukes (2005: 12) has more recently commented that it was a mistake to define power by 'saying that A exercises power over B when A affects B in a manner contrary to B's interests'. Instead, Lukes argues that power is a capacity rather than the exercise of that capacity (which may never even have to be exercised). Instead, power is the imposition of internal constraints with those subject to such constraints being 'led to acquire beliefs and form desires that result in their consenting or adapting to being dominated, in coercive or non-coercive settings' (Lukes, 2005: 13).

Lukes' (2005) approach to the third dimensions of power recalls Bourdieu's (2000 [1997]) ideas with respect to how the maintenance of 'habitus' appeal to the workings of power, 'leading those subject to it to see their condition as "natural" and even to value it, and to fail to recognize the sources of their desires and beliefs' (Lukes, 2005: 13). Such domination is, according to Bourdieu,

> exerted not in the pure logic of knowing consciousness but through the schemes of perception, appreciation and action that are constitutive of habitus and which, below the level of the decisions of consciousness and the controls of the will, set up a cognitive relationship that is profoundly obscure to itself. (Bourdieu, 2000 [1997]: 37)

However, perhaps much more significantly for the study of power in a tourism context, Lukes (2005: 91) has downplayed the potential explanatory contribution of Foucault to understanding the structure of power, noting that his idea of power, 'in its non-overstated and non-exaggerated form, is simply this: that if power is to be effective, those subject to it must be rendered susceptible to its effects'.

According to Foucault (1987: 11), 'the subject constitutes himself in an active fashion, by the practices of self'. These practices are not invented by the individuals, but are derived from 'patterns that he finds in the culture and which are proposed, suggested and imposed on him by his culture, his society and his social group'. Yet, as Lukes (2005) suggests, what is therefore so supposedly radical about the Foucauldian notion of power (referred to by Digesser (1992) as the fourth face of power)? Instead, it restates some elementary sociological understandings that would be known to any first-year student.

> Individuals are socialized: they are oriented to roles and practices that are culturally and socially given; they internalise these and may experience them as freely chosen; indeed, their freedom may, as Durkheim liked to say, be the fruit of regulation – the outcomes of discipline and controls. Of course, it restates these truths in a distinctively Foucauldian way... (Lukes, 2005: 97)

Indeed, Lukes goes on to note that Foucault's notion that power is 'productive' through the social construction of subjects therefore actually makes no sense in terms of understanding how the various modern forms of power actually succeed or fail in securing compliance. Nevertheless, it is apparent that Foucault's writings have had a wide impact, seen throughout many of the chapters in the present volume, leading Lukes (2005: 98) to suggest that 'Foucault's writings thereby themselves exhibit an interesting kind of power: the power of seduction'.

Nevertheless, we are still left with the task of determining whether the shaping and imposition of values constitute domination and the denial of real interests: a challenging task given the extent to which it is difficult to write of real interests given that people's interests can be varied, conflicting and situational. Luke's response is that what counts as 'real interests' should be interpreted as 'a function of one's explanatory purpose, framework and methods, which in turn have to be justified' (Lukes, 2005: 148). However, such an approach is not to deny that 'false consciousness' does not exist, but to suggest that rather than being understood as some sort of assertion that one has privileged access to truths denied to others, it needs to be understood as the cognitive power to mislead (Lukes, 2005). Such power can be witnessed in censorship and disinformation as well as in the denial of other ways of thinking and doing. However, such power is not all-embracing, cracks do appear in walls, and power's third dimension is partial at best over time, as power does meet resistance.

Conclusions

This chapter has sought to stress that power is a multi-layered concept. Arguably, the overt exercise of power, and even non-decision making, are not terribly controversial concepts in the realm of tourism studies, although they are surprisingly little examined, especially in the realm of tourism policy and planning. Indeed, it may be wondered if the relative lack of studies of overt exercise of power by tourism academics is because of concerns over being seen to challenge the role of government and industry interests in decision making, who also fund the tourism programmes within which some tourism researchers are situated.

As has been suggested elsewhere (Hall, 2007a), the value of a Lukesian approach to power is highlighted in the multi-layering of observations of power occurring in the three faces or dimensions of power that provides an empirical strength often missing in Foucauldian analyses that, while acknowledging the role of structural dominance, often fails to record the actions of individual actors in relation to specific issues and interests (i.e. Cheong & Miller, 2000). Similarly, there is a substantial amount of such writing in tourism where authors have exhorted the notion of a tourist gaze without examining the concepts of power and knowledge on which it is grounded and given little thought to the role that individual actors play with respect to power relations from a decision and non-decision-making perspective. Moreover, there is also insufficient attention to how academia and tourism research is caught within its own web of power relations 'which plays out through a particular industrial actor-network of academic knowledge production, circulation and reception' (Gibson & Klocker, 2004: 425). Such criticisms are important because of the way in which academic research, including the anthropology of tourism, is often woven into the manufacture of commodified culture and heritage for tourism purposes. While being critical, many academic actors are at the same time beneficiaries of the power structure of the academy and of the culture industry. Yet the implications of such relations are little reflected on.

Culture and heritage provides a useful setting in which to investigate the third and other dimensions of power. Institutional representations and reconstructions of culture and heritage are often not fully inclusive. Some historical actors are included while others are left out. As noted above, in any set of institutional arrangements, what is left out is often more important than what is left in. Particular representations reveal themselves to the tourist through museums, historic houses, historic monuments and markers, guided tours, public spaces, heritage precincts and landscapes in a manner that may act to legitimate current social and political structures. As Norkunas (1993: 5) recognised, 'The public would accept as "true" history that is written, exhibited, or otherwise publicly sanctioned. What is often less obvious to the public is that the writing or the exhibition itself is reflective of a particular ideology'. Perhaps in the same way, we should also encourage more reflection on the writing of the academy and the extent to which it reflects its own set of power relationships.

The exercise of cultural power in terms of tourism and the subsequent occupation of cultural space may have a profound effect on identity. But often the exercise of power with respect to tourism and culture is based purely on the actions of one set of political actors towards another (Hall, 2007b). However, the exercise of such power is variable over space and at different scales of analysis. Tourism at times silences voices. While the Romantic representation of culture still holds sway in much tourism promotion, substantial fragmentation is occurring at the local level with respect to tourism. Such fragmentation and subsequent reassertion of identity by groups who had previously been denied their cultural voice is related to broader processes of globalisation and localisation within the cultural and economic dynamic of late capitalism and the neo-liberal project. Tourism is an integral component of this dynamic and, as such, provides a mechanism to articulate previously silenced voices as it does to silence them. It is too simplistic to suppose that unwilling and willing compliance to domination are mutually exclusive, one can consent to power and resent the mode of its exercise. Given the binary relationship of tourism to culture, identity and representation, it would therefore seem essential to look beyond tourism in order to understand the nexus of interests, values and power.

Notes
1. Often not realised by students of tourism is that one of Urry's first books was a co-edited book on power in Britain (Urry & Wakeford, 1973).

References

Anderson, P. (1976–1977) The antinominies of Antonio Gramsci. *New Left Review* 100, 5–78.
Ateljevic, I. and Doorne, S. (2002) Representing New Zealand: Tourism imagery and ideology. *Annals of Tourism Research* 29 (3), 648–667.
Bachrach, P. (1967) *The Theory of Democratic Elitism: A Critique.* Boston, MA: Little, Brown.
Bachrach, P. and Baratz, M.S. (1962) Two faces of power. *American Political Science Review* 56, 947–952.
Bachrach, P. and Baratz, M.S. (1970) *Power and Poverty: Theory and Practice.* Oxford: Oxford University Press.
Bourdieu, P. (2000 [1997]) *Pascalian Meditations* (R. Nice, trans.). Stanford, CA: Stanford University Press.
Cheong, S. and Miller, M. (2000) Power and tourism: A Foucauldian observation. *Annals of Tourism Research* 27 (2), 371–390.
Church, A. and Coles, T. (eds) (2007) *Tourism, Power and Space.* London: Routledge.
Crenson, M. (1971) *The Un-Politics of Air Pollution: A Study of Non-Decision-making in the Cities.* Baltimore, MD: John Hopkins Press.
Dahl, R.A. (1961) *Who Governs? Democracy and Power in an American City.* New Haven, CT: Yale University Press.
Digesser, P. (1992) The fourth face of power. *Journal of Politics* 54 (4), 977–1007.
Domhoff, G.W. (2007) C. Wright Mills, Floyd Hunter, and 50 years of power structure research. *Michigan Sociological Review* 21, 1–54.
Doorne, S. (1998) Power, participation and perception: An insider's perspective on the politics of the Wellington waterfront redevelopment. *Current Issues in Tourism* 1 (2), 129–166.
Elections BC (2002) *Referendum 2002: Report of the Chief Electoral Office on the Treaty Negotiations Referendum.* Victoria, British Columbia: Elections BC.
Fernia, J. (1981) *Gramsci's Political Thought: Hegemony, Consciousness and the Revolutionary Process.* Oxford: Clarendon Press.

Foucault, M. (1984) The subject and power. In B. Wallis (ed.) *Art After Modernism: Rethinking Representation* (pp. 417–432). New York: New York Museum of Contemporary Art.

Foucault, M. (1987) The ethic of care for the self as a practice of freedom: An interview with Michel Foucault on 20 January 1984. In J. Bernauer and D. Rasmussen (eds) *The Final Foucault* (pp. 1–20). Cambridge, MA: MIT Press.

Gallie, W.B. (1955–1956) Essentially contested concepts. *Proceedings of the Aristotelian Society* 56, 167–198.

Gibson, C. and Klocker, N. (2004) Academic publishing as 'creative' industry, and recent discourse of 'creative economies': Some critical reflections. *Area* 36 (4), 423–434.

Hall, C.M. (1994) *Tourism and Politics: Power, Policy and Place.* London: John Wiley.

Hall, C.M. (2007a) Tourism, governance and the (mis-)location of power. In A. Church and T. Coles (eds) *Tourism, Power and Space* (pp. 247–269). London: Routledge.

Hall, C.M. (2007b) Politics, power and indigenous tourism. In R. Butler and T. Hinch (eds) *Tourism and Indigenous Peoples: Issues and Implications* (pp. 305–318). Oxford: Butterworth-Heinemann.

Hall, C.M. (2008) *Tourism Planning* (2nd edn). Harlow: Prentice-Hall.

Hall, C.M. and Jenkins, J.M. (1995) *Tourism and Public Policy.* London: Routledge.

Holmes, S. (1988) Gag rules or the politics of omission. In J. Elster and R. Slagstad (eds) *Constitutionalism and Democracy* (pp. 19–58). Cambridge: Cambridge University Press.

Hunter, F. (1953) *Community Power Structure: A Study of Decision Makers.* Chapel Hill, NC: University of North Carolina Press.

Hyland, J.L. (1995) *Democratic Theory: The Philosophical Foundations.* Manchester: Manchester University Press.

Langton, M. and Palmer, L. (2003) Modern agreement making and indigenous people in Australia: Issues and trends. *Australian Indigenous Law Reporter* 8 (1), 1–31.

Lasswell, H.D. (1936) *Politics: Who Gets, What, When, How?* New York: McGraw-Hill.

Lukes, S. (1974) *Power: A Radical View.* London: MacMillan.

Lukes, S. (2005) *Power: A Radical View* (2nd edn). Basingstoke: Palgrave Macmillan.

MacCannell, D. (1976) *The Tourist: A New Theory of the Leisure Class.* New York: Schocken Books.

MacCannell, D. (1999) *The Tourist: A New Theory of the Leisure Class* (3rd edn). Berkeley, CA: University of California Press.

Millar, C. and Aiken, D. (1995) Conflict resolution in aquaculture: A matter of trust. In A. Boghen (ed.) *Coldwater Aquaculture in Atlantic Canada* (2nd edn; pp. 617–645). Moncton, New Brunswick: The Canadian Institute for Research on Regional Development.

Mills, C.W. (1956) *The Power Elite.* New York: Oxford University Press.

Ministry of Advanced Education and Treaty Negotiations Office (2004) News Release: B.C. boosts tourism training, jobs for first nations, Ministry of Advanced Education and Treaty Negotiations Office, Vancouver, March 12.

Mokken, R.J. and Stokman, F.N. (1976) Power and influence as political phenomena. In B. Barry (ed.) *Power and Political Theory: Some European Perspectives* (pp. 33–54). London: John Wiley.

Murphy, P. (1985) *Tourism: A Community Approach*. New York: Methuen.

Norkunas, M.K. (1993) *The Politics of Memory: Tourism, History, and Ethnicity in Monterey, California*. Albany, NY: State University of New York Press.

Rojek, C. (1985) *Capitalism and Leisure Theory*. London: Routledge.

Rojek, C. (1995) *Decentring Leisure: Rethinking Leisure Theory*. London: Sage.

Rossiter, D. and Wood, P. (2005) Fantastic topographies: Neo-liberal responses to Aboriginal land claims in British Columbia. *The Canadian Geographer* 49 (4), 352–366.

Russell, B. (1938) *Power: A New Social Analysis*. London: Allen and Unwin.

Schattsneider, E. (1960) *Semi-sovereign People: A Realists View of Democracy in America*. New York: Holt, Rinehart and Wilson.

Urry, J. (1990) *The Tourist Gaze: Leisure and Travel in Contemporary Societies*. London: Sage.

Urry, J. and Wakeford, J. (eds) (1973) *Power In Britain: Sociological Readings*. London: Heinemann Educational.

Watson, G.L. and Kopachevsky, J.P. (1994) Interpretations of tourism as a commodity. *Annals of Tourism Research* 21 (3), 643–660.

Wolf, E.R. (1999) *Envisioning Power: Ideologies of Dominance and Crisis*. Berkeley, CA: University of California Press.

Index

214

CPSIA information can be obtained at www.ICGtesting.com
Printed in the USA
BVOW04s0154070214

343987BV00006B/91/P

9 781845 411244